Chemical Applications of Molecular Modelling

University of
Hertfordshire
Library

Reference only

Chemical Applications of Molecular Modelling

Jonathan M. Goodman
*Royal Society University Research Fellow, Department of
Chemistry, University of Cambridge, UK, and Fellow of Clare
College Cambridge*

ROYAL SOCIETY OF CHEMISTRY

ISBN 0-85404-579-1

A catalogue record for this book is available from the British Library

Published by The Royal Society of Chemistry,
Thomas Graham House, Science Park, Milton Road, Cambridge CB4 4WF, UK

For further information see our web site at www.rsc.org

Typeset by Paston Press Ltd, Loddon, Norfolk
Printed by Redwood Books Ltd, Trowbridge, Wiltshire

Preface

Organic chemistry has a reputation for being a difficult subject. As computers become an important part of everyday life, so molecular modelling is affecting all parts of chemistry, and it is particularly useful in organic chemistry. This book is an attempt to demystify molecular modelling, so that the non-specialist can appreciate the power and the limitations of the computational tools which are available. I hope that this might make organic chemistry easier!

This book is designed to be read by anyone interested in organic chemistry and molecular modelling, and not just by those who already have access to a molecular modelling system. The examples sections are presented as series of questions that can be investigated using molecular modelling, but all the examples are then explained in detail, so it is not necessary to have done the appropriate calculations in order to follow these sections.

This book is intended to provide an introduction to molecular modelling for the experimental chemist, and to show how these techniques can aid the experimental chemist throughout the synthetic process, from the selection of synthetic targets to the analysis and differentiation of stereoisomers. It is aimed at the pragmatic scientist, who wants to use molecular modelling and is faced with the questions: What can we find out? How reliable is it? What is a reasonable question to ask?

Molecular modelling does not provide chemistry's final solution, but it can contribute to many important chemical problems. I hope this book will enable more people to benefit from the use of this technique.

A web page associated with this book is available on:

http://www.chemsoc.org/gateway/chembyte/Goodman.htm

Any errata will be notified on this web page.

Acknowledgements

This book has evolved from courses on molecular modelling which have been run in Cambridge by J. G. 'Andy' Vinter and myself for several years. I am very grateful to Andy for his advice and input to this book. He is responsible for many of the ideas, and none of the mistakes! I would also like to thank Ian Paterson and Scott Kahn, who first introduced me to molecular modelling, and Clark Still, under whose guidance I was able to learn more about this subject.

Many people have provided comments and suggestions on the molecular modelling courses and draft versions of this book, including Mark Gardner, Keith Trollope, Mark Mackey, Paul Wolstenholme-Hogg, Thomas Trieselmann, Andrew Dominey, Michael Kranz, David Riddick, Tomoko Masaike and Jack Bikker. I am grateful for all their help.

Jonathan Goodman
Cambridge, December 1997

Contents

CHAPTER 1

Introduction

1.1 WHAT IS A MOLECULE?

A molecule may be defined as a collection of atoms linked by bonds. This simple idea, that molecules are three-dimensional structures which obey the rules of civil engineering, can help to account for a wide range of phenomena, even though quantum theory has shown that this is a simplification. Bonds can bend and stretch and this can be seen in *infrared* spectra. If the bending and stretching is taken to an extreme, the molecule will break. Bonds are able to twist like axles and these torsional movements can be stimulated by radiation in the microwave region. This aspect of molecular architecture means that a molecule which has freely rotating bonds can take up many shapes (conformations) without breaking any of its bonds.

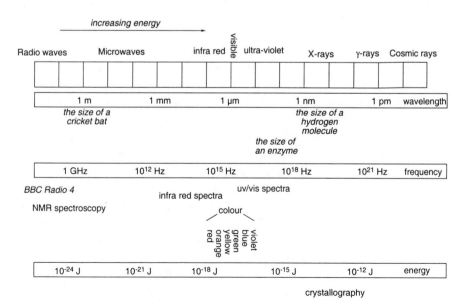

Figure 1.1 *The electromagnetic spectrum*

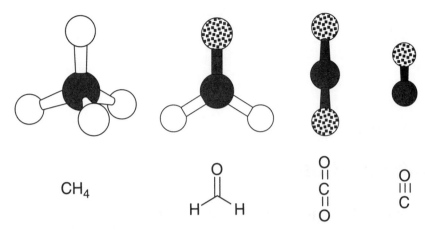

Figure 1.2 *Methane, formaldehyde, carbon dioxide, carbon monoxide*

Carbon atoms can form bonds to one, two, three or four neighbouring atoms (Figure 1.2), and these neighbours tend to be as far from each other as possible. A carbon attached to four atoms, such as methane, will have a tetrahedral geometry, and the angle between the bonds will be 109.47°. With three neighbours, the bonds can be 120° apart by forming a flat structure, as they do in formaldehyde. With only two neighbours, a linear structure is possible. Other atoms will also adopt these shapes, although some, such as oxygen, have lone pairs, which will take up the position of a bond without forming a connection to another atom. Thus water, H_2O, is not a linear molecule, but more like a tetrahedron with two arms missing, because oxygen has two lone pairs. Most of the structures in organic chemistry can be built up from these simple components. The heavier elements can form more complicated shapes, for which it is rather harder to build simple mechanical models.

Atoms seem to have a definite size. For convenience, an atom can be regarded as a sphere with a particular radius, called the van der Waals radius (after Johannes van der Waals, Leiden, 1873, who first noticed these interactions), which varies from element to element and with the charge on the atom. Two atoms will not want to approach each other closer than the sum of their van der Waals radii, unless they are connected by a chemical bond, and this effect is called van der Waals repulsion. However, there is also van der Waals attraction: atoms do want to be close to each other. If two atoms are separated by the sum of their van der Waals radii, it will take a lot of energy to push them closer together. It will also require energy to pull them apart. This short-range attraction is very important. The way in which energy changes with separation is sketched in Figure 1.3. The energy climbs rapidly as the distance between two atoms becomes very close, but also increases when

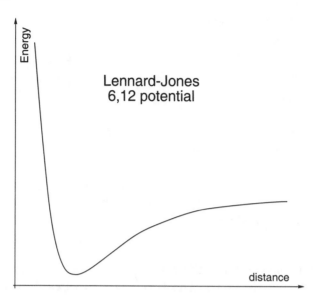

Figure 1.3 *A Lennard-Jones 6,12 potential*

the atoms are separated beyond an optimal distance. Figure 1.3 shows a Lennard-Jones 6,12 potential, but the shape of the curve is rather similar for the other potential energy functions which are also used.

Individual atoms within a molecule are not usually electrically neutral, although the molecule as a whole may be. There are, therefore, Coulombic forces pushing and pulling on the atoms. This electrical force is a long-range effect, unlike the van der Waals interaction, and can be very influential; ion-pair and salt-bridge formation are electrostatically driven, as are hydrogen bonds.

A molecule can be visualised as a flexible scaffold with a specific directional geometry to each joint, based on the tetrahedral angle of 109.47° or the trigonal angle of 120.0°. The struts bend and flex very quickly. They rotate on their axes less quickly and the whole molecule can rotate and move. Each joint is surrounded by a large sphere, which prevents it getting too close to any other joint. Less easy to fit into the engineering analogy is the idea that the atoms can attract and repel each other electrostatically and the movement and redistribution of charge occurs at rates which far exceed the *infra-red* frequencies and range from the visible region down to the far *ultra-violet*.

Using this simple engineering analogy, a *molecular mechanics* model of a system of atoms can be constructed. An early example of this was Barton's study of decalin (Barton, 1948). Decalin, which has ten saturated carbon atoms arranged in two rings (Figure 1.4), has two diastereoisomers: one with the ring junction hydrogens on opposite sides

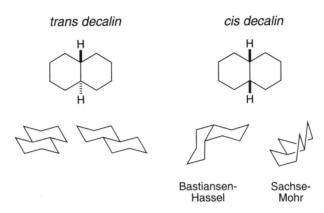

Figure 1.4 Trans- *and* cis-*decalin*

of the rings (*trans*) and the other with the ring junction hydrogens on the same side of the ring (*cis*). The two 3-D structures drawn for *trans*-decalin in Figure 1.4 are identical, just viewed from different angles. The structure for *cis*-decalin is harder to draw, and there were two possible structures: the Bastiansen–Hassel structure and the Sachse–Mohr structure, which were different and both, apparently, reasonable.

By constructing a molecular mechanics model of the system, Barton was able to show that the Bastiansen–Hassel structure has a lower strain energy than the Sachse–Mohr structure, and so is more likely to represent the true conformation of the system. The calculation also showed that *trans*-decalin was lower in energy than *cis*-decalin, which was consistent with measurements of the heats of combustion of the two compounds.

There are many groups of atoms which often occur together. For example, carbon and oxygen are often doubly-bonded to each other, and this is called a carbonyl group. An oxygen bonded to a hydrogen forms a hydroxyl group. Adding an hydroxyl group to the carbon of a carbonyl creates a new group called a carboxyl, which is an acidic moiety because of its ability to lose a proton (a hydrogen cation) and become a carboxylate anion. Is it possible to treat a molecule as a collection of groups whose constituent parts define the whole molecule? Quantum mechanics shows that bonds form as a consequence of electron redistribution from every atom in the molecule. This implies that every atom will be different and that chemistry cannot be analysed on an atomic or group basis. However, experiment reassures us that, in practice, this can be a useful approach. Infra-red spectroscopy is, on the one hand, a unique fingerprint of the molecule and highlights its individuality. On the other hand, the wavelength range over which which a particular group will respond is usually similar regardless of the molecule which contains it. Nuclear magnetic resonance spectroscopy is a method of observing the effect of electron density around atomic nuclei. The chemical shift of a nucleus will be high if it is in an electron-deficient environment and low if it is shielded by lots of electrons. Groups usually appear within narrow ranges of chemical shift regardless of the overall molecular structure. The consideration of functional groups is, therefore, a useful way to analyse molecules. In the same way, molecular modelling is a useful tool which can be used to obtain valuable information about molecules, despite its approximations.

1.2 MOLECULAR CHANGE AND ENERGY

1.2.1 Thermodynamics

Molecules are constantly moving, changing shape and, occasionally, losing and gaining pieces through rearrangement or reaction. Some shapes (from now on, the word *conformation* will be used instead of shape to refer to the different ways in which a molecule can twist around its bonds) are lower in energy than others, because the bonds are less stretched, or they avoid unfavourable van der Waals interactions. In an isolated system, a molecule cannot lose energy, so the lower energy conformations must be moving more rapidly, balancing their decreased strain energy with greater kinetic energy. This does not seem to correspond to a scaffolding model, because most scaffolding does not move very fast, if all is well. If scaffolding collapses, however, it loses its potential energy, and this is absorbed by the environment as vibrations of various sorts. This can be simulated in a molecular model. The energy can be removed from a molecule, driving it to a low energy conformation. This process is often called *energy minimisation*, and is described in detail in Chapter 3.

If an isolated molecule cannot lose energy, why should it prefer some conformations to others? It could be argued that it has no reason to prefer any particular conformation over any of the others as it has the same total energy for all of them. This conclusion is incorrect. If a scaffolding structure falls down, no-one would expect it spontaneously to reassemble itself, even though all of the energy necessary for this is still in the environment. Nothing can change from a disordered state to a more ordered state, unless there is some external source of energy. This fundamental observation is a statement of the second law of thermodynamics, which can also be phrased 'entropy increases' where entropy is a measure of disorder. This also explains why an isolated molecule will prefer certain conformations. Its total energy does not change, but its entropy can increase. It will, therefore, choose the conformations which give the system the highest entropy, and these correspond to the states with low strain energy, because this maximises the amount of movement of the molecule. The word 'energy' has been used to mean the total of the molecule's strain energy and kinetic energy. This is not the same as the molecule's *free energy*, which is given the symbol *G*. A change in free energy is given by:

$$\Delta G = \Delta H - T\Delta S \tag{1.1}$$

where ΔG is a change in free energy, ΔH is a change in enthalpy, which roughly corresponds to the energy that has been discussed up to now, *T* is the temperature, which must be measured in Kelvin, not degrees centigrade, and ΔS which is a change in entropy. This equation makes it clear that a change in strain energy is not the same as a change in free energy. The change in free energy is a useful quantity because it can be related to the equilibrium constant between different states:

$$-\Delta G = RT \ln K \tag{1.2}$$

In this equation, *R* is the gas constant, which has a value of $8.314\,\mathrm{J\,K^{-1}\,mol^{-1}}$ (there is a list of useful constants in Appendix A.1), *T* is the temperature, in Kelvin, and *K* is an equilibrium constant, the ratio of two states. Accounts of the origin of Equations 1.1 and 1.2 can be found in most physical chemistry textbooks. A few are listed under Further Reading.

Consider the general situation of a state A in equilibrium with a state B (Figure 1.5 illustrates these as different conformations of cyclohexane, but it could equally well be a light switch in the on and off positions, or anything else which can be in two states). The overall free energy difference between A and B is ΔG. The equilibrium constant, *K*, is the ratio of the amount of the system in state A to the amount in state B (The concentrations of A and B are usually written [A] and [B], respectively, so

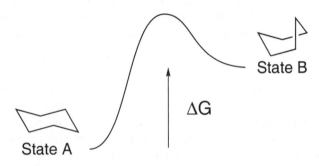

Figure 1.5 *Two conformations of cyclohexane*

$K = \frac{[A]}{[B]}$). If there are the same amounts of state A and state B, then the equilibrium constant is one, so ΔG is zero (since $\ln 1 = 0$). If there is more of state A than state B, then ΔG must be positive, as illustrated. For a particular free energy difference, the ratio of the amount of state A to the amount of state B will depend on the temperature. Equation 1.2 tells us that at high temperatures, there will be more of state B, the higher energy state, than at low temperatures. This fits in with common sense. In order to increase the amount of the high energy state, give the system more energy (heat). However, the equation also tells us that there can never be more of state B than state A, even if the temperature tends to infinite values (this assumes that ΔG does not change with temperature, which is often a good approximation). Figure 1.5 also shows that there is an energy barrier to overcome, if state B is to turn into state A. This does not affect the position of the equilibrium (the thermodynamic properties of the system), only the rate at which equilibrium is attained (the kinetic properties, discussed in the next section).

The energy difference between the boat and chair forms of cyclohexane is approximately $20 \, kJ \, mol^{-1}$ (in fact, this is an internal energy difference, calculated by a molecular modelling method, and not a free energy difference. In this case, because the entropies of the two states are not likely to be dramatically different, and because the relative values, not the absolute values, of energy are being used, this is a reasonable approximation.) The ratio between the two forms at room temperature (say 300 K) will be about 3000 to 1 in favour of A. This preference will increase at lower temperatures. This means that an energy difference of only $20 \, kJ \, mol^{-1}$ is enough to give almost entirely state A, and almost none of state B. The strength of a carbon–carbon bond is about $350 \, kJ \, mol^{-1}$, so they are effectively unbreakable at room temperature. An energy difference of $2 \, kJ \, mol^{-1}$ would give a ratio of about two to one at room temperature. A chart plotting the effects of different energy gaps at various temperatures is available in Appendix A.3.

1.2.2 Kinetics

How long will a system take to come to equilibrium? This question requires a knowledge of the kinetics of the system, and so the energy barrier between the different states, $\Delta G^{\ddagger}$ (Figure 1.6). The Eyring equation, which is derived in many physical chemistry text books, can be used to obtain information about this sort of system. A more simplistic analysis, however, can be used to give a very approximate idea of how fast reactions are likely to be. The chance that state B will make a successful assault on the energy barrier $\Delta G^{\ddagger}$ can be estimated by

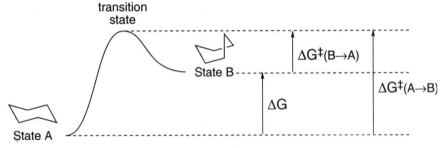

Figure 1.6 *Energy barriers for changing the conformation of cyclohexane*

assuming an equilibrium between the transition state and state B. The 'concentration' of the transition state can, therefore, be found from Equation 1.2:

$$[\text{Transition state}] = [\text{state B}] \exp\left(\frac{-\Delta G^{\ddagger}}{RT}\right) \qquad (1.3)$$

To get an expression for the rate of the reaction, it is necessary to estimate the number of times a second that state B attempts to turn into state A. An extremely crude approximation estimates the energy of state B as kT, where k is the Boltzmann constant and T is the temperature in Kelvin. This energy is associated with a frequency, which can be found by dividing the energy by Planck's constant, h. At room temperature, this gives a frequency of about 10^{12} s^{-1} (1 THz). Multiplying this number by the concentration of the transition state gives an expression for the rate of any reaction (Equation 1.4). Despite the extreme approximations in this approach, it gives answers which are correct, within a couple of orders of magnitude, for many systems.

$$\text{Rate of many reactions is very approximately equal to} \frac{kT}{h}\exp\left(\frac{-\Delta G^{\ddagger}}{RT}\right) \qquad (1.4)$$

In order to have a check on this expression, it is useful to consider the Arrhenius equation (Equation 1.5), which states that the the rate constant for a reaction is equal to a constant A multiplied by e to the power of the activation energy divided by the gas constant and the temperature (in Kelvin, of course). Typical values for the constant A are between 10^{10} s^{-1} and 10^{15} s^{-1}, for first order reactions. The value of 10^{12}, which was calculated for Equation 1.4 as the pre-exponential factor at room temperature, falls neatly into the middle of this range. The energy ($E^{\ddagger}$) in the Arrhenius equation does not correspond precisely to the free energy of activation ($\Delta G^{\ddagger}$) in Equation 1.4, but at this level of approximation the distinction is not important.

$$\text{Rate constant} = A \exp\left(\frac{-E^{\ddagger}}{RT}\right) \tag{1.5}$$

Equation 1.4 can be used to provide very crude estimates of the rate of various processes at room temperature. The barrier for rotating the two methyl groups of ethane, relative to each other, is about $10\,\text{kJ}\,\text{mol}^{-1}$. The groups will, therefore, spin about 10^{10} times every second. The barrier for the conversion of chair cyclohexane to boat cyclohexane is about $40\,\text{kJ}\,\text{mol}^{-1}$, which means the conversion will occur about one hundred thousand times a second. A process with an energy barrier of $100\,\text{kJ}\,\text{mol}^{-1}$, however, will only occur about twice a week. Graphs of Equation 1.4 are given in Appendix A.4.

1.3 THE CONTROL OF THERMODYNAMICS AND KINETICS

1.3.1 A Real Example

The carbonyl group of the molecule in Figure 1.7 squeezes through the large chain as it moves from boat to chair 8 times a second ($\Delta G^{\ddagger} = 65\,\text{kJ}\,\text{mol}^{-1}$, Vinter and Hoffmann, 1973, 1974). The ground-state free energy difference between the chair the boat is virtually zero in a non-polar solvent so there is a fifty-fifty mixture of the two conformations at equilibrium.

It is interesting to examine what happens when one methylene is added or removed from the large ring in the molecule (Figure 1.8). When the ring is increased to thirteen carbons, the interconversion energy of activation decreases from $65\,\text{kJ}\,\text{mol}^{-1}$ to less than $40\,\text{kJ}\,\text{mol}^{-1}$. As for the twelve-membered ring, the free energy difference ($3\,\text{kJ}\,\text{mol}^{-1}$) between the chair and the boat forms is small enough that both can be seen by NMR at room temperature.

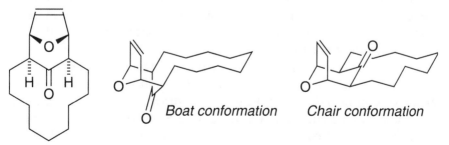

Boat conformation   *Chair conformation*

Figure 1.7

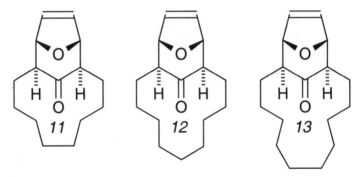

Figure 1.8

However, on removing one carbon to form the eleven-membered ring, the activation energy for interconversion increases to more than 90 kJ mol^{-1}, which virtually stops the process at room temperature. The energy of the chair conformation also rises relative to the boat, and so its concentration decreases, to the extent that it disappears from the NMR spectrum.

If the boat form of the molecule is separated and its double bond reduced to a single bond, the methylene hydrogens now interfere with the bridging ring in the axial position (in the boat) and increase the free energy of this conformation relative to the chair. If the temperature is raised to 100 °C, it is observed by NMR that all the boat conformation changes into the chair form. There can be no going back because ΔG is too high.

1.3.2 Conformation Searching

The example in Figure 1.8 illustrates an important problem with molecular modelling. Minimising the strain energy (which will be labelled E_{MM}) of a structure, so that all the bonds are about the right length, all the angles are about right and there are no bad van der Waals interactions, can give a reasonable looking structure. Techniques to perform this minimisation process are discussed in Chapter 3. However, there is no guarantee that this will give the structure that the molecule wants to adopt. In the example above, minimisation may give the boat form, when the chair form is preferred.

To take a simpler example, consider butane, a linear hydrocarbon with four carbons in its chain (Figure 1.9). It is natural to draw it in the extended conformation, but it can also be drawn in a twisted conformation. The twisted conformation is about 3.5 kJ mol^{-1} higher in energy than the extended conformation. If a random conformation of butane were constructed, and its energy minimised, only one structure would be

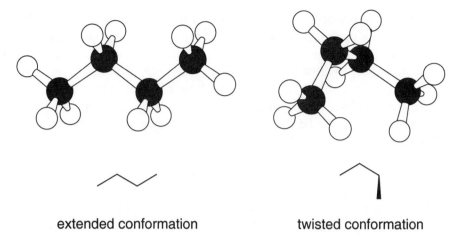

extended conformation   twisted conformation

Figure 1.9 *Butane*

obtained, and it could be either the extended or the twisted conformation. Which one is 'correct'?

The answer is that they are both 'correct' and a complete picture of the conformational properties of butane can only be found by considering both of them. This problem is far more dramatic for larger molecules. Instead of two conformations which need to be considered, there may be hundreds, thousands or millions!

This problem can be overcome by *conformation searching*, which explores all the possible conformations of a molecule. This process can be very time consuming, and many methods have been developed, which are described in detail in Chapter 4. However the problem is addressed, it is certain to be time consuming, and this is why it is necessary not only to be able to calculate the energy of a molecule and to minimise this quantity, but to be able to do this quickly, so that the many minimisations required for conformation searching can all be performed in a reasonable length of time.

1.3.3 Molecular Orbitals and Quantum Mechanics

Purely mechanical models of molecular structure can produce very useful structural information. However, they take no account of electrons, which play an important role in chemistry! Molecular orbital theory, which is discussed in Chapter 6, can be used to provide a much more precise description of molecular structure than a mechanical model, and can also provide precise information about electronic properties. Why bother with molecular mechanics, if it is possible to solve Schrödinger's equation and so gain a complete description of molecules?

Whilst it is possible to solve Schrödinger's equation for molecular systems, some approximations must be introduced, and the process is extremely time consuming. A molecular mechanics calculation on a small molecule, such as decalin, may well take less than a second, because there are only a few atoms and bonds to consider. A molecular orbital calculation on the same system must take account of all the atomic nuclei and all the electrons, and consider how each one interacts with every other one. Inevitably, it will take much longer. To calculate the properties of a molecule twice as big as decalin, molecular mechanics will take approximately four times as long to complete the calculation (the calculations scale as N^2). A molecular orbital calculation on a system twice the size will probably take more than sixteen times as long (the calculations scale as N^4 or, for some precise methods, as N^8). Things get more difficult very rapidly as molecules get larger. When the conformation searching problem is combined with this, it is clear that molecular orbital theory is simply too difficult and requires too much computer power for many problems.

1.4 HOW DO MOLECULES MOVE?

The bending and stretching of bonds takes between ten and a hundred femtoseconds. Every bond in the molecule is moving in this way, and so they jostle against each other in condensed phases. Studies of water/deuterium hydroxide mixtures early this century demonstrated that a molecule of water moved its own diameter though the liquid every thousand vibrations (Orr and Butler, 1935). This would correspond to about 20 m s^{-1} if the molecule were to move in a straight line at this rate. Because every molecule will be jostling against others and changing direction, its rate of progression over larger distances will be very much slower than this.

As well as moving around and vibrating, molecules will also twist around their bonds, continually changing from one conformation to another. It is possible to simulate how these movements occur, by giving a computational model of a molecule some energy, randomly spread around the atoms, and calculating how the molecule moves by solving Newton's laws of motion. This is called molecular dynamics, and is discussed in Chapter 5. Since this can, in principle, give a description of how a molecule will change between all of its conformations, you might hope that the problem of defining conformational space was solved.

In practice, this is not the case, because the calculations require an enormous amount of computer time. In order to solve the coupled differential equations which arise from Newton's laws of motion, it is necessary to move forward along the path they define in very small steps. If the fastest vibrations take about ten femtoseconds (10^{-14} s), it is not possible to take steps much larger than a single femtosecond. The barriers to rotation around bonds are usually between $10 \, \text{kJ} \, \text{mol}^{-1}$ and $40 \, \text{kJ} \, \text{mol}^{-1}$, so it may be estimated from Equation 1.3 that a molecule may only take something of the magnitude of 10^{-10} s to explore a single bond's rotation, provided it has a low barrier to rotation. This does not sound very long, but it must be simulated in 1 fs steps, of which a hundred thousand will be required. Molecules of greater complexity will take much longer. Experimental data suggest that some proteins take around about a second to fold from their denatured state to their native conformation. In principle this could be simulated with a molecular dynamics calculation, but it would take 10^{15} steps, each of which would require an energy calculation on the whole protein. A powerful computer might be able to do a hundred such calculations every second. If so, it would be able to simulate the folding of a protein in about three hundred thousand years. Computer power needs to increase by a factor of at least a million for this sort of calculation to be feasible, and this is likely to take some time, even at the current extraordinary rate of growth. Fortunately, it is possible to get useful information from shorter calculations, but care must be taken to choose questions which can be answered.

1.5 SUMMARY

The energy of a molecule is a useful property for investigating the behaviour of molecules. This energy can be approximated by using simple mechanical models for molecules, and this provides a rapid method for its calculation, provided sufficient parameters are available, that is, we know how long every bond is, how strong it is and so on. Minimisation of the molecular mechanics energy of a system gives unstrained conformations, but will not necessarily find the lowest energy conformations. For this, some form of conformation searching or molecular dynamics simulation is required, and these may be very time consuming. A molecule's energy may also be calculated using molecular orbital theory, which is more accurate but much slower than molecular mechanics. If a conformation search is required, then molecular orbital calculations may be too slow, even for quite small molecules. If it is necessary to include solvent molecules in a calculation, which is often the case in order to calculate properties which can be measured by organic chemists, a molecular orbital treatment may be impossible.

Molecular mechanics is empirical, approximate and rather less than general. It is, however, the only technique which is able to tackle many of the important questions in chemistry. The fact that it works at all is tribute to the empirical approach and to the dedication of those who have collected data and tested parameters with painstaking care.

CHAPTER 2

Force Fields

2.1 ENERGY CALCULATION

2.1.1 What is Being Calculated?

It is possible to build mechanical models of molecules, and these can give useful information about their structure. Metal and plastic models have been available for a long time, but it is now possible to build such models in computers, and so undertake quantitative analyses of their properties. What is it possible to calculate from such a molecular mechanics model?

The quantity that is most interesting for many experiments is the free energy, G, but this is rather hard to calculate directly, because it includes a measure of the entropy of the molecule. A molecular mechanics model will readily give a value for 'energy' simply by adding up the strain in all of the bonds and the van der Waals and Coulombic interactions of all of the atoms. This quantity can be called E_{MM}, the molecular mechanics energy (it is also called the 'steric energy'). This is not G, so what is it?

The property that E_{MM} most closely mirrors is the internal energy of a molecule, which is given the symbol U. It is not easy to relate a zero point of E_{MM} to a zero point of U, so in practice it is usual to look at differences in energy, ΔE_{MM} and ΔU, rather than absolute energies. This has the additional advantage that errors in the calculation may cancel.

The standard equations of thermodynamics (for detailed explanations, refer to the physical chemistry books listed under Further Reading), relate U to other quantities. Enthalpy, H, is defined as:

$$H = U + PV \qquad (2.1)$$

Here P refers to pressure and V to volume, neither of which are easy to define in a molecular mechanics model. Considering only changes in enthalpy simplifies the picture. At constant pressure ($\Delta P = 0$):

$$\Delta H = \Delta U + P\Delta V \qquad (2.2)$$

For a simple molecular mechanics model there is no external pressure, so:

$$\Delta H \approx \Delta U \approx \Delta E_{MM} \tag{2.3}$$

A change in free energy, ΔG, is related to a change in enthalpy, ΔH, by the equation which was given in the last chapter as Equation 1.1, but is so important that it is repeated here:

$$\Delta G = \Delta H - T\Delta S \tag{2.4}$$

In this equation, T is the absolute temperature (measured in Kelvin), and ΔS is a change in entropy. If there is reason to believe that the change in entropy for a process is likely to be small ($\Delta S \approx 0$), then a change in molecular mechanics energy, ΔE_{MM}, may be a reasonable approximation for a change in free energy. This approximation is often used, sometimes without due caution.

2.2 WHAT IS IN A FORCE FIELD?

The molecular mechanics energy, E_{MM}, is made up of a number of components. The energy in every bond, E_{bonds}, is added to the energy in every angle, E_{angles}, and to the energy of all the van der Waals interactions, E_{vdw}. The earliest quantitative molecular mechanics models used only these terms (Hill, 1946; Westheimer and Mayer, 1946; Barton, 1948). It soon became clear that a term for torsion angles, $E_{torsion}$, was also required in order to explain many properties. For molecules with electronegative groups, charge interactions, E_{charge}, must also be included. Many different groups have developed force fields, and they all follow this scheme, Equation 2.5, although most have additional terms as well, which will be referred to as $E_{miscellaneous}$. These methods were sufficiently mature for a review to be necessary in 1956 (Westheimer, 1956), and the advantages of 'machine computation' were advocated by Hendrickson in 1961 (Hendrickson, 1961). The description in this chapter is based mainly on MM2, a very widely used force field (Allinger, 1977), but the general pattern is similar for most force fields.

$$E_{MM} = E_{bonds} + E_{angles} + E_{vdw} + E_{torsion} + E_{charge} + E_{miscellaneous} \tag{2.5}$$

2.2.1 Bond Energy: E_{bonds}

The first component of E_{MM} to consider is bond stretching. The simplest model for this is to treat bonds as springs, and this is what is usually done (Figure 2.1). The energy of the bond is simply a constant multiplied by

Figure 2.1 *A chemical bond treated as a spring*

the square of the displacement from the equilibrium position, which is a simple spring, as described by Hooke's law, and gives an harmonic energy curve.

$$E_{bonds} = k_1 (l - l_0)^2 \qquad (2.6)$$

This means that the energy goes up whether the bond is pushed or pulled, and continues to go up however far the stretching continues. This is not what happens with real molecules, of course. If a bond is stretched sufficiently far, it will break. The way in which energy varies with distortion is sketched as the solid line in Figure 2.2, with the Hooke's law approximation shown as a dashed line. The solid line closely resembles the curve for van der Waals interactions, Figure 1.3. (Curves of this shape occur so often that they are sometimes called 'the chemical curve'.) The shape of the curve is rather similar to the van der Waals interaction, but the scale is very different. It only takes a few kJ mol^{-1} to break up a van der Waals complex. Breaking a chemical bond may take a hundred times as much energy.

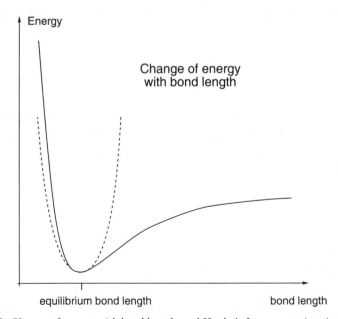

Figure 2.2 *Change of energy with bond length, and Hooke's Law approximation (dashed)*

The harmonic approximation to the true energy profile does not look very good. It only fits well close to the equilibrium bond length. This emphasises a limitation of most molecular mechanics force fields: extreme structures are not treated very precisely. Fortunately, this is not too much of a problem, because covalent bonds are very strong compared with the other forces affecting molecules, so they rarely change very much from their equilibrium bond length. Force fields may modify Equation 2.6 with another term proportional to the cube of the change in bond length. This can make the right hand side of the curve go down, fitting the true energy profile more precisely. This also means that extreme displacements are actually favoured, so the program must check carefully that this does not happen.

This simple model is useless, of course, unless the equilibrium bond length, l_0, and the bond stretching constant, k_1, are known for every bond in a molecule. The main difficulty in creating a new force field is building up a reliable and consistent database of such quantities, based on experimental values. X-ray diffraction provides a huge amount of data on bond lengths, and infra-red spectroscopy provides information about bond strengths, because absorptions in the infra-red region tend to correspond to bonds stretching. It is possible, therefore, to assemble such databases. However, there is an element of choice. Is a carbon–carbon single bond going to be exactly the same length if it is in a molecule of ethane, a piece of diamond, a strand of DNA or an enzyme? Molecular mechanics would not be useful if separate values for l_0 and k_1 were needed for carbon–carbon bonds in every different situation. Fortunately, the experimental values are similar for these very different molecules.

2.2.2 Bond Angle Energy: E_{angles}

Bond angles are next to consider (Figure 2.3). The simplest model is chosen once more: the energy of bending a bond angle is taken to be proportional to the square of the displacement from equilibrium (Equation 2.7).

$$E_{angles} = k_\theta (\theta - \theta_0)^2 \tag{2.7}$$

As with bond lengths, this expression is only realistic for small displacements from equilibrium, but this is not too bad, because bond angles cannot change very much. As before, some force fields modify this expression to improve its behaviour in extreme cases. Values for the parameters can be obtained from X-ray diffraction and infra-red spectroscopy, as before.

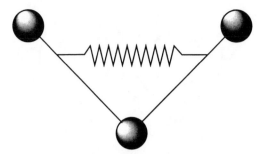

Figure 2.3 *Angle bending*

2.2.3 van der Waals Interactions: E_{vdw}

Atoms cannot get too close together (Figure 2.4). If they approach beyond a particular distance, and if a bond does not form, the energy of interaction goes up very rapidly. Although atoms are just clouds of electrons surrounding a tiny nucleus, they behave as if they have a quite definite size, and a measure of this is the van der Waals radius. A table of values is available in Appendix A.5. The size of atoms can be estimated using X-ray diffraction data from crystals, and cross-sectional areas from the kinetic theory of gases (Bondi, 1964). Different studies tend to give slightly different answers, and the radii depend on the oxidation state and environment of each atom. However, the values do not vary by a large amount, so it is reasonable to take an average value.

Energy increases very rapidly as atoms become too close. The precise relationship between separation and repulsive energy is not clear, and may be estimated by various expressions. The Buckingham potential uses an exponential form, and this is the one favoured by Allinger for MM2. The Lennard-Jones potential uses a $1/r^{12}$ term, and various other expressions have been used, giving rather similar results.

Although atoms cannot get too close, they do like to be adjacent (Figure 2.5). The reason for their mutual attraction is an induced-dipole interaction. An uncharged isolated atom will not have a dipole, on average, but because the electrons are moving around the nucleus, it is

Figure 2.4 *Atoms getting too close*

Figure 2.5 *Atoms at optimal proximity*

possible for an instantaneous dipole to exist. This will induce a dipole on any nearby atom, and the net effect will be attraction. This is sometimes called the London force or a dispersion interaction. The net effect is proportional to $1/r^6$. This makes the calculation of a Lennard-Jones 6,12 potential rather easy to carry out, since $1/r^{12} = (1/r^6)^2$. This means that it is easy to calculate the $1/r^{12}$ term once the $1/r^6$ calculation has been carried out. Such considerations were very important when the calculations were done by hand, or with mechanical calculating machines. It is still important, even though computers are now very powerful, because the limiting factor for many studies is the amount of computer time required.

Non-bonding interactions are usually only calculated for atoms which are separated by three or more bonds. This is because the close interactions are handled by bond stretching and bond angle bending. If van der Waals interactions were included for atoms which are bonded to each other, the value of E_{MM} would be enormous, because atoms which are bonded can be much closer than the sum of their van der Waals radii. The same is true for 1,3 interactions.

Values for the van der Waals radii of atoms are fairly easy to estimate from a variety of experiments. The 'hardness' of atoms is much more difficult. The development of a force field is not a deterministic process, with each parameter fixed by experiment. Different choices of parameters may give reasonable results, provided all the different atoms' parameters are chosen consistently. For example, it is possible to estimate the hardness of atoms (Hill, 1948). These values can give reasonable results when used in a force field. However, much larger, but softer, atoms (Bartell, 1960) give rather similar results.

There is a particular problem with describing hydrogen. The nucleus of an isolated atom will be in the centre of its electrons. This is not necessarily the case if the atom has formed bonds to other atoms. If the bonds are formed symmetrically, as they are for the carbon atom of methane, the nucleus will remain central. If the bonds are not symmetrical, as for the hydrogen atoms of methane, then it may not remain central. This is a particularly pronounced effect for hydrogen, because its orbitals are smaller than those of other atoms. Force fields sometimes allow for this by shortening all bonds to hydrogen atoms when the van der Waals interactions are calculated, thus allowing for the polarisation of electrons. For example, when the MM2 force field calculates van der Waals interactions, it treats the hydrogen nucleus as if it were only 91% of the way along each C—H bond, rather than at the end as might be expected. This leads to a better description of molecular structures.

2.2.4 Torsion Angle Energy: $E_{torsion}$

Molecules can rotate around single bonds, and there is an energy barrier for such rotations. This energy barrier is implicit in the contributions to E_{MM} which have already been described. Ethane prefers to be in a staggered conformation, and rotating around the carbon–carbon bond requires energy (Figure 2.6). This barrier can be attributed to the van der Waals repulsion of the hydrogen atoms squeezing past each other, because hydrogen atoms are certainly large enough to 'touch' in this sort of situation. The values of the parameters in the force field determine the barrier. Unfortunately, they do not do a very good job, in all cases. A choice of parameters which works well for ethane may work less well for cyclohexane. This problem is overcome by introducing additional parameters for this interaction. It is not easy to simply add a spring (Figure 2.7) as before, and a truncated Fourier series is used instead, based on the torsion angle defined by the four atoms involved, designated θ in Equation 2.8.

$$E_{torsion} = \frac{V_1}{2}(1 + \cos\theta) + \frac{V_2}{2}(1 - \cos 2\theta) + \frac{V_3}{2}(1 + \cos 3\theta) \qquad (2.8)$$

The torsional profile for ethane is sketched in Figure 2.8. The distance from the bottom of the troughs to the top of the peaks is about $10\,\text{kJ}\,\text{mol}^{-1}$, the barrier for rotating a methyl group. A graph of this shape can be generated using cosines, and an expression is given in Equation 2.8. Each of the parameters V_1, V_2 and V_3 is divided by two,

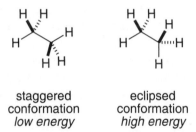

staggered
conformation
low energy

eclipsed
conformation
high energy

Figure 2.6 *Ethane*

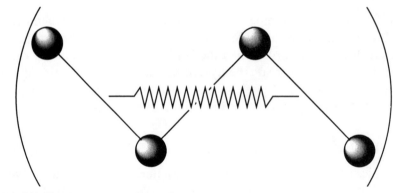

Figure 2.7 *Constraining a torsion angle*

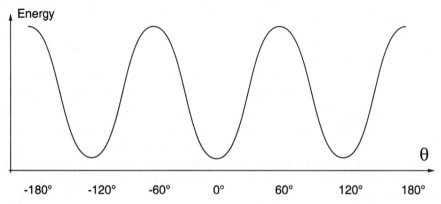

Figure 2.8 *Torsion profile for rotating methyl groups of ethane. The barrier to rotation is about 10 kJ mol^{-1}*

because the cosine function ranges from plus one to minus one. Rotation barriers can be measured by microwave spectroscopy and by NMR experiments.

When deciding on van der Waals parameters, force field designers have a choice between small hard atoms and large squashy atoms. Torsion terms introduce a further element of choice into force field design. What is the balance between the torsion barrier due to van der Waals interactions and the torsion barrier due to expressions such as Equation 2.8? There is no simple answer to this question, as there appears to be a partial redundancy in the parameters. This means that the choice of torsion parameters depends on the choice of van der Waals parameters, and also the values chosen to describe bond lengths and bond angles. A parameter cannot, therefore, simply be transferred from one force field to another, because it will not necessarily be balanced with the choices made in the design of the rest of the force field.

2.2.4.1 Improper Torsion. Torsion terms are also used to keep sp^2 atoms flat. The tetrahedral shape of an sp^3 atom is maintained by the bond angle terms, but this is not enough to keep an sp^2 atom flat, as a very small change in bond angle can lead to significant pyramidalisation. An additional energy term, usually called an *improper torsion* term, is often introduced. Experimental data show that sp^2 atoms, such as the carbons in double bonds, may often show significant distortions from planarity. The improper torsion term gives such atoms a preference for planarity, but allows distortion where necessary.

2.2.5 Charge–Charge Interactions: E_{charge}

The first force fields were designed with hydrocarbons in mind, and charge–charge interactions were not of great significance. If E_{MM} is simply regarded as the sum of the terms described above ($E_{MM} = E_{bonds} + E_{angles} + E_{vdw} + E_{torsion}$) a reasonable model results for simple systems. The success of force fields at describing the properties of hydrocarbons led to more ambitious attempts to create force fields which could calculate the properties of a wider range of molecules. Charge interactions now became important. For example, a carbonyl group's electron density is polarised towards the oxygen, and so the energy of interaction of two carbonyls will be different if they are aligned or opposed (Figure 2.9). Aligning the groups brings two partial negative charges close to each other, which is unfavourable compared with the opposite arrangement, which pairs partial positive and partial negative charges.

The simplest way to quantify this effect is to assign a partial charge to every atom and to use Coulomb's Law (Equation 2.9) to calculate the energy of interaction (MM2 uses an alternative approach, based on bond dipoles).

$$E_{charge} = \frac{1}{4\pi\varepsilon}\frac{q_1 q_2}{r} \tag{2.9}$$

The force between charged particles is proportional to $1/r^2$ and so the energy of interaction is proportional to $1/r$. The partial charges on two

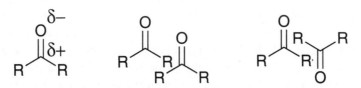

Figure 2.9 *Charge–charge interactions in carbonyl groups*

different atoms are represented by q_1 and q_2. The ε in the equation is the permittivity of the medium, and this has the value of 8.854×10^{-12} C^2 J^{-1} m^{-1} in a vacuum (the permittivity of a vacuum is given the symbol ε_0). The distances involved on a molecular level are of the order of Ångstroms (10^{-10} metres), and the partial charges are usually less than the charge of an electron (1.60×10^{-19} C). Will multiplying these very small values give an answer of a reasonable magnitude? For example, what is the energy of interaction of two points each with one tenth of the charge of an electron, 3 Å apart? This is the sort of interaction which often occurs in molecular mechanics models. The separation, r, is 3×10^{-10} m, and the charges, q_1 and q_2, are both 1.60×10^{-20} C. E_{charge}, therefore, is 7.7×10^{-21} J. This is an extremely small figure, but is the energy per pair of charges, and not per mole, as all previous energies were expressed. Multiplying by Avogadro's number ($N_A = 6.02 \times 10^{23}$ mol^{-1}) gives an energy of interaction of $4.6\,kJ\,mol^{-1}$, which is the same magnitude as the energies for the other components of E_{MM}.

The total value of E_{charge} can only be found by adding up the results of Equation 2.9 for all pairs of atoms. Thus, if a molecule has N atoms, there will be N^2 interactions which must be considered. This will be difficult for larger molecules. The time required to calculate the total E_{bonds}, E_{angles} and $E_{torsion}$ will increase approximately linearly with the size of the molecule, so for larger molecules most of the difficulty in the calculating the energy of the molecule will be in summing the very many non-bonded interactions. The task is sometimes simplified by using a cut-off distance: if atoms are separated by more than a certain distance, they are assumed to have no interaction. This works well for E_{vdw}, because the attractive interaction falls off very rapidly with distance, but this can work less well for charge–charge interactions, as there can be a significant interaction over quite large distances.

2.2.5.1 Permittivity.   The correct value to use for ε, the permittivity, has been the subject of much discussion. It is convenient to write:

$$\varepsilon = \varepsilon_0\, \varepsilon_r \qquad\qquad (2.10)$$

In this expression, ε_0 is the permittivity of a vacuum, and ε_r is a dimensionless quantity called the relative permittivity. In a vacuum, then, $\varepsilon_r = 1$. For water, ε_r has been measured as 80. For ethanol it is 24 and for benzene 2.3. If the charge–charge interaction of a molecule is being calculated, what is the right value to use? At an atomic level, it could be argued that atoms exist in a vacuum, so ε_r should be set equal to one. This appears to overestimate charge–charge interactions, because there may be other atoms between the two whose energy is being calculated. Some force fields arbitrarily set ε_r to a larger value. A commonly used approximation is to set ε_r to the value of the separation of the charges, measured in Ångstrom. This is referred to as a distance dependent dielectric, and has the result that the energy of interaction is proportional to $1/r^2$, not $1/r$. Different force field developers have made a variety of choices for their molecular mechanics models.

2.2.5.2 Determination of Partial Charges. There is another serious difficulty with parameterising this part of a force field. How is it possible to decide on the partial charge for each atom? Since atoms are not point charges, but positively charged nuclei surrounded by clouds of electron density, answering the question is equivalent to dividing the clouds of electrons between the nuclei. The electron distribution can be found from X-ray diffraction, but the experiments are very hard to do. Molecular orbital theory can be used to calculate the electron distribution, for simple molecules, and various schemes are available to divide the electron density between atoms. This is discussed in more detail in Chapter 6.

An alternative approach is to calculate partial charges by consideration of the electronegativity of atoms. The idea of electronegativity, the power of an atom to attract electrons, was introduced by Pauling (Pauling, 1939). The electronegativity of an atom will naturally depend on its charge. This suggests a way of calculating the charges on atoms in a molecule (Sanderson, 1951): the charges on all the atoms in a molecule are varied until they all have the same electronegativity. This should give a charge distribution which reflects the electronegativity of the elements. This idea has been implemented by Gasteiger (1980).

However the partial charges on a molecule are parameterised, they can be verified by measurement of the dipole moment of the whole molecule, which is a relatively easy measurement to make. This is only a rather crude measure of the electronic distribution of a molecule. The values chosen for partial charges vary widely between force fields.

2.2.5.3 Beyond Atom-Centred Point Charges. Putting charges on the nuclei of all the atoms in a molecule is not a very precise way of modelling the charge distribution. This is a particular problem for molecules with lone pairs. One way to improve the electrostatic model is to incorporate extra 'atoms' in the molecule, which help represent the lone pairs on a system. Figure 2.10 shows dimethyl ether. The extra 'atoms' on the central oxygen are lone pairs—small additional atoms with negative partial charges.

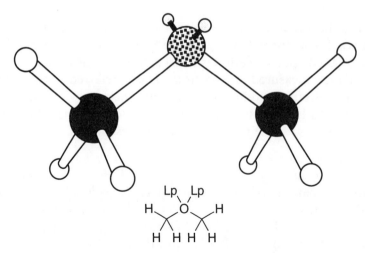

Figure 2.10 *Dimethyl ether, showing additional 'atoms' to represent the lone pairs on oxygen*

2.2.6 Miscellaneous Interactions: $E_{miscellaneous}$

Most force fields also have some other terms, called 'cross terms'. If a bond is stretched, it may be easier to bend the associated bond angles. If a bond angle is opened out, the barrier for a torsion rotation may be reduced. These effects can be incorporated into force fields, by including expressions depending on the pairs of terms which might interact. This makes the force field much more complicated, but should allow it to fit better to the data that were used to parameterise it. It is not clear, however, whether such extra terms help the molecular mechanics model to calculate the properties of molecules which were not included in the parameterisation data.

2.3 HOW ARE MOLECULES DESCRIBED?

In order to do a molecular mechanics calculation, it is necessary to have a precise description of a molecule. Many different file formats are used for this, but they all contain coordinates for all the atoms, a list of the bonds connecting the atoms, and a description of each atom. The last of these sounds as if it should be simple, because the most important property of an atom is its atomic number, which determines which element it is. However, things are not so simple. Carbon, atomic number 6, can be tetrahedral, trigonal planar or linear, and the force field must be told which shape to make any particular atom. As a result, there are usually many more atom types than there are elements. A molecule may also be simplified by combining elements. A methyl group is more or less spherical, so why not represent it by a single large 'atom' instead of a carbon and three hydrogens? This should make a calculation go more rapidly. This approach is often used, especially for very large molecules, but the increased speed is offset by lower accuracy.

For example, a file describing propene (CH_3—CH=CH_2) is shown in Table 2.1, in the MacroModel format (Mohamedi *et al.*, 1990).

The first line of the file just contains the number of atoms in the structure, 9 in this example. The next nine lines each describe an atom. The first column is the atom type, which in this format is given by a number. An sp^3 carbon has atom type 3 and an sp^2 carbon has atom type 2. The first three atoms in the file are the three carbons of propene, and the remaining lines, with atom type 41, are all of the hydrogens.

The next columns describe the bonds in the molecule. The first atom, the sp^3 carbon, is connected to another carbon and three hydrogen atoms, each with a single bond. The second and third columns '2 1' mean that this first atom is connected to the second atom in the file, with a single bond. Looking on to the second atom, there is a '1 1' entry, which shows that the second atom is also connected to the first, with a single bond. Columns four to thirteen contain the rest of the bond information. The second atom is connected to the third with a double bond, so '3 2' occurs in its list of bonds, and '2 2' occurs in the list for the third atom. This format limits the total number of bonds to any particular atom to

Table 2.1

9								
3	2 1	4 1	5 1	6 1	0 0	0 0	6.592566	5.936225	0.000000
2	1 1	3 2	7 1	0 0	0 0	0 0	7.974285	6.532524	0.000000
2	2 2	8 1	9 1	0 0	0 0	0 0	9.101203	5.806678	0.000000
41	1 1	0 0	0 0	0 0	0 0	0 0	6.612274	4.822587	0.000000
41	1 1	0 0	0 0	0 0	0 0	0 0	6.032619	6.269237	0.903381
41	1 1	0 0	0 0	0 0	0 0	0 0	6.032619	6.269237	−0.903381
41	2 1	0 0	0 0	0 0	0 0	0 0	8.044492	7.634272	0.000000
41	3 1	0 0	0 0	0 0	0 0	0 0	9.074581	4.705400	0.000000
41	3 1	0 0	0 0	0 0	0 0	0 0	10.090425	6.292901	0.000000

six. This is not a severe restriction, since this format is usually only used for 'organic' molecules, for which seven or more bonds to a single atom are almost unknown.

Columns fourteen, fifteen and sixteen contain the cartesian coordinates of each atom. The units of these number are Ångstroms. Macro-Model files can also have more columns, with additional information about atomic charge, colour, and so on being stored to the right of the cartesian coordinates.

There are so many file formats in use that it is not possible to list them all. They usually contain the information described above, using many different styles and conventions, and may also contain a wide range of other information.

When a molecular mechanics program is given a file such as this, it can make a list of all the interactions that are present. In this example, there are C—H bonds, a C—C bond and a C=C bond. There are H—C—H, H—C—C, H—C=C, and C—C=C angles, and there are H—C—C—H, H—C=C—H, H—C—C=C, and H—C=C—C torsion angles to consider. The data for all of these must be looked up in a force field, which contains a long list of such interactions. It is immediately clear that there are a great many interactions, even for such a simple molecule. The situation is further complicated by the distinction between a H—C(sp^3) bond and a H—C(sp^2) bond, and the corresponding combinations for all the angles. Whilst there are a great many terms to consider, each individual term is fairly straightforward, so all that needs to be done is to calculate every one and add them all up. This would be extremely dull to do by hand, but is no problem for a computer. Thus, an energy can be found for this geometry of propene, provided all of the necessary interactions had been included in the force field.

Propene can also be described using united atoms, that is, with the hydrogen atoms included in the carbon atoms (Table 2.2).

Now there are only three atoms in the molecule. Atom type 6 is a

Table 2.2

3									
6	2 1	0 0	0 0	0 0	0 0	0 0	5.792312	5.615086	0.000000
7	1 1	3 2	0 0	0 0	0 0	0 0	5.979548	4.104647	0.000000
8	2 2	0 0	0 0	0 0	0 0	0 0	7.186972	3.530453	0.000000

methyl group, atom type 7 is an sp^2 CH group and atom type 8 is an sp^2 CH_2 group. The energy calculation will now be very much faster, but the reliability of the energy obtained will be lower.

2.3.1 Water

Water is such an important molecule that special descriptions of it have been developed. Jorgensen has developed the TIP3P and the TIP4P models (Jorgensen *et al.*, 1983), which are widely used as they can provide a good approximation to the properties of liquid water at reasonable computational expense.

2.4 TRANSFERABILITY

Force fields comprise long lists of parameters for the bond lengths, strengths, angles and so on. These parameters, which are derived by careful comparison with experiments and other calculations, should form a consistent set. However, different force fields have different choices for the size and hardness of atoms, for the use of additional 'atoms' representing lone pairs and for the presence or absence of cross terms. It is not, in general, possible to take a parameter from one force field and put it into another. Parameters are not transferable between force fields. Force fields have often been designed to tackle particular problems, such as proteins or strained hydrocarbons. There is no 'best' force field for all problems, and the choice depends on the particular system of interest, and the choice between speed and accuracy which must often be made.

Are force fields transferable between molecules? This is a much more crucial question. Can a force field that was developed using data from a particular set of molecules be used to get useful information about different molecules? The answer can only be found by trying the force field out and comparing the results with new experimental data. Some force fields have been tested with a wide range of molecules, and there can be more confidence in these than in force fields which have only been used in a few specific studies. A list of force fields is given in Appendix A.7. It seems to be the case that it is possible to use force fields to analyse and quantify the properties of a wide range of molecules, and so they are extremely useful tools for organic chemists. In principle, a force field can provide all possible information about the conformational properties of a molecule. The only difficulty lies in extracting this information in a usable form. The next three chapters provide ways of doing this.

2.5 KEY POINTS

- Force fields provide an empirical way of calculating E_{MM}, which is sufficiently accurate to be useful.
- Force field calculations are very fast.
- Force fields work best for 'organic' systems. In general, there are no good parameters for metal–carbon bonds.
- Non-bonded interactions tend to be the weakest aspect of many force fields.
- Parameters should not be transferred between different force fields.
- Adding parameters to force fields is possible, but hard work (see Appendix A.12).

Minimisation

3.1 INTRODUCTION

Chapter 2 described how it was possible to design a force field to calculate a value for E_{MM} for a molecule, from a file containing a list of atoms, bonds and coordinates for the molecule. Where can such files be obtained? There are databases which contain many molecular structures, and these can be a useful starting point. It is more common, however, simply to sketch the molecule on a computer screen, to get an approximate geometry and then to minimise its energy.

The process of minimisation is illustrated for a diatomic molecule in Figure 3.1. The initial position A can move to a lower energy by making a small change to the geometry of the system. It can be moved either left or right. In this case, moving to the left lowers the energy, so the initial position can be forgotten, and the process repeated from B, then C, then D, then E. The process is continued until all possible small changes to the structure increase the energy of the structure. Once at position E, all possible small changes will increase the energy of the system. In one dimension, as illustrated, this looks very simple. However, a molecule will have $3N - 6$ degrees of freedom, where N is the number of atoms in the molecule, and so the process becomes rather complicated. It is obvious from the diagram where the minimum energy point is, but plotting the curve required the calculation of the energy at a large number of points, and drawing a smooth line between them. In a simple case like this one, the process does not take very long, and it may even be possible to find an analytical expression for the position of the lowest energy point of the function. If, however, a typical expression for E_{MM}, the molecular mechanics energy of a molecule, is used, then the calculation of each point will take a significant amount of time, because every point requires the calculation and summation of all the components of E_{MM}. It is not possible to get an analytical expression for the lowest point on the surface in such a situation. The only way to find a minimum energy structure is to start at a random or a promising position, and roll downhill.

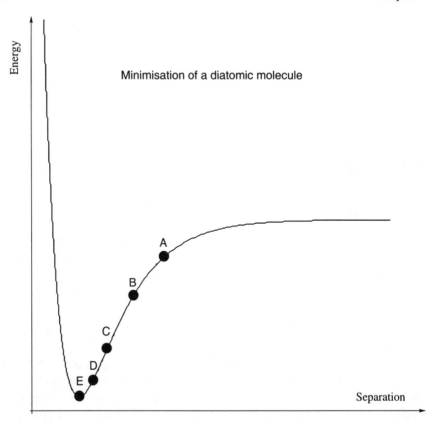

Figure 3.1 *Minimisation*

An important effect of minimisation is that structures are 'tidied up'. If cyclohexane is sketched on a computer screen, then the bond lengths will vary, the bond angles will not be very close to the tetrahedral angle and the whole thing will be a bit messy. If a minimisation algorithm is applied, a sensible looking structure for cyclohexane should be produced.

This is illustrated for cyclohexane in Figure 3.2. A very distorted structure (1) is minimised towards a neat chair-shaped cyclohexane (12). In the first structure, the bond angles and bond lengths are both very distorted. In structure 2, the bond lengths look almost right, although the bond angles still need some correction. By structure 6, the geometry looks almost right, and structures 7 to 12 are all very similar.

This illustrates some important points about minimisation. It would appear that the program regards the problem with the bond length as the most important one, and this is corrected quickly. The bond angles are then put right, without affecting the bond lengths. Finally, small adjustments are made to optimise the structure. In fact, the program

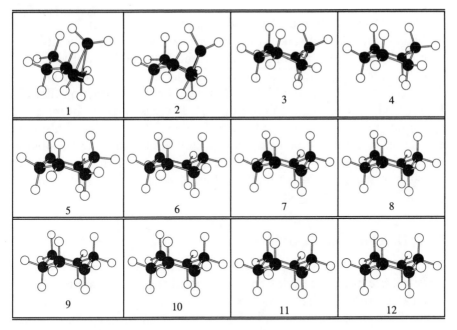

Figure 3.2 *Cartoon strip of cyclohexane minimising*

does not divide up the problem in this way. It just tries to minimise the molecular mechanics energy (E_{MM}, described in the previous chapter), as rapidly as possible. A small distortion in the bond length will give a large change in E_{MM}, so the first steps of minimisation will probably concentrate on the bond length terms. The last steps produce very little change, and it may seem from this figure that the minimisation could perfectly well have stopped at structure 10, or even at structure 6. However, the minimiser will have its own criterion for when to stop, and in order to get accurate energies, minimisation must continue after obvious motion has ceased.

In order to write a computer program which will minimise E_{MM}, several questions must be addressed:

(i) Which way do we go?
(ii) How far can we go?
(iii) How do we know when we have arrived?

3.1.1 Which Way Do We Go?

This may seem ridiculously simple, but there are two things to notice. First, it is necessary to decide in which direction to move. In one dimension, there is only one variable, the bond length, and so the choice is whether to increase it or decrease it (going right or left in Figure 3.1). In higher dimensional space, this becomes a much more difficult question, because there is no longer one variable but hundreds of variables. How to alter all of them to move downhill best? The most efficient direction will probably depend on thinking about the direction that has been travelled to get to the present point. There is no point in retracing steps, nor in tacking back and forth. Imagine walking on a hillside in thick fog. It is clear which way is downhill, and it may be possible to remember the area that has just been walked through, but it is not possible to see if moving in this direction is actually the best way to get downhill as fast as possible.

3.1.2 How Far Can We Go?

To get downhill as fast as possible, it is necessary to take as few steps as possible, because after each step the energy of the new geometry must be calculated, and a decision must be made about which direction to step next. If the steps are too small, there will be a great many steps, which will be time consuming. If the steps are too large, the minimum energy point may be passed, and time may be wasted by retracing steps.

3.1.3 How Do We Know When We Have Arrived?

This looks obvious, from a glance at the figure for a one-dimensional minimisation. We have arrived when we reach the lowest energy point. However, a computer program cannot see the whole curve as we can in the diagram, but only knows about the few points on the curve that have been calculated. One way to be absolutely sure is to make every possible small change to the structure that may be the minimum energy structure, and to check that the energy increases in each case. In order to do this, it is necessary to define what a small change is. Because we know we are dealing with molecules, this is not too hard. A small fraction of an Ångstrom will be small enough. Checking all possible changes, however, may be very time consuming in a molecule which contains dozens of atoms.

   Minimisation algorithms may require the calculation of the first derivatives of the energy and the second derivatives of the energy with respect to the coordinates of each atom. If the potential energy surface is flat, then all of the first derivatives will be zero. The second derivatives will, in general, not be zero, and they can give information about the shape of the potential energy surface around the stationary point. The matrix of second derivatives (which will have $9 \times N^2$ entries, where N is the number of atoms in a molecule, because each entry corresponds to two sorts of movement) is called the Hessian matrix, and analysis of this can distinguish between minima, maxima and saddle points.

3.2 MINIMISATION ALGORITHMS

These three problems are not trivial ones. There are many different algorithms available to solve the problem, which may be thought to indicate there is no single best way of doing it. The best answer in a particular case depends on the characteristics of the molecule being studied and the force field being used, as well as the precision which is required in the answer. Some common minimisation algorithms are outlined below. All of them are widely used, and any one of them (except for steepest descents) may be the best for a particular problem.

The choice between different minimisation methods may often not make much difference. Most people will have a favourite, and will use no other unless something goes wrong. However, most minimisation methods will have problems in particular situations, and it can be worthwhile trying more than one if problems arise. Derivative methods, which decide on the direction to move from the slope of the potential energy surface, are generally very effective. Sometimes, however, it is particularly difficult to calculate the slope of a potential energy surface, and in such situations a non-derivative method may be preferable.

3.2.1 Non-derivative Methods

These are methods that do not require the calculation of the derivative (the slope) of the potential energy surface. As a result, they are easier to write programs for, but are generally less efficient. The most commonly used non-derivative method is *Simplex Minimisation.*

In one dimension this method is easy to describe. Choose an arbitrary step size. Take a step of this size from your starting point. If the energy goes up, try a step in the other direction. Go on taking steps, until you get to a point where taking a step forwards and also a step backwards will raise the energy of the system. Now is the time to take smaller steps. Continue the process until the steps are very small. Very small, in this context, would usually mean so small as to make no very significant difference to a molecule, which may be about 0.01 Å.

In two dimensions, the process is similar. The minimisation begins with three points, and the energy is calculated at each of them. In Figure 3.3, the points A, B and C are chosen initially. One of them must have the highest energy, in this case A. The figure is therefore reflected through the line BC, to construct a new point D, which is away from the highest energy point A. The process is repeated, constructing new points away

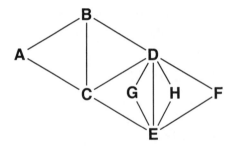

Figure 3.3 *A simplex minimisation in two dimensions*

from the highest energy point in the current triangle. In this way, the triangle 'walks' downhill. The algorithm usually has some additional complexity. If the triangle is marching steadily in a particular direction, then larger steps can be taken.

When the triangle has marched downhill towards the minimum, it may step over the minimum. In the diagram, C is higher in energy than D or E, so point F is constructed. Clearly now, however, F is higher in energy that D and E. Rather than just return to point C and oscillate forever, the step size is reduced, and G, and then H are tested. The search would continue until some criterion of step size was so small that no progress was being made in any direction.

An analogous process can occur in more than two dimensions, by the construction of a tetrahedron or a higher dimensional figure, as appropriate for the problem. This figure would wander around the potential energy surface, changing shape as it went, until it came to rest at the minimum.

This way of minimising functions is quite straightforward to program, and is useful for problems in which the derivative of the function is not available. For example, it is used in the optimisation of chicken feed mixtures. The relative ease with which this algorithm can be turned into a computer program has found it support in some areas of computational chemistry.

3.2.2 Derivative Methods

It is usually possible to calculate the slope of a potential surface from the potential energy function. In the one-dimensional case, this tells us how steep the line is, and which way is down. In a two-dimensional case, it tells us which direction goes downhill most steeply. In a multidimensional case, this is not easy to imagine, but the equations will give the

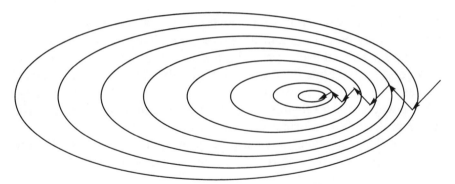

Figure 3.4 *A two-dimensional minimisation, which needed to think more about the direction to take*

answer. There are four classes of methods which use derivative information:

   (i) Steepest descents
  (ii) Newton Raphson
 (iii) Conjugate gradients
 (iv) Variable metrics

3.2.2.1 Steepest Descents. The simplest way to use the derivative information is simply to go downhill in the direction of the steepest slope. The method is called *steepest descents*, and it sounds good. It provides an answer to the question 'Which way should we go?' but not to the question 'How far should we go?' This second question may be answered by using *line searching*. A one-dimensional minimisation is much easier than a multidimensional one, so once the steepest descent direction has been chosen, we can do a one-dimensional minimisation along that direction. This is called line searching and it provides an answer to the question 'How far should we go?'

Unfortunately, this combination of methods does not work particularly well. If you imagine walking down a sloping valley, you do not want to tack from side to side, but just walk down the floor of the valley. Steepest descents will keep you tacking, because each new step will be perpendicular to the last. This is illustrated in Figure 3.4. Steepest descents is not usually the best choice of minimisation method.

3.2.2.2 Newton Raphson. Newton Raphson methods use both the gradient of the potential energy surface and also its curvature. This is a very effective way of minimising energies. It is rather time consuming to calculate the curvature of the potential energy surface precisely, and so approximations are often made to speed up the process. This is called block-diagonal Newton Raphson or truncated Newton Raphson. These methods work better on large molecules. For small systems full Newton Raphson may be the method of choice. The algorithm spends a lot of time thinking about each step that it makes, but usually decides to make very good steps. It has the drawback that it can find transition states as well as true minima. A transition state is an area of the potential energy surface where the gradient is zero, as for a minimum, but small changes in the molecule's conformation will not always raise the energy of the molecule. This drawback may reasonably be regarded as a useful feature, as transition states may be very interesting areas of the potential energy surface.

3.2.2.3 Conjugate Gradients and Variable Metrics. Conjugate gradient methods and variable metric methods are entirely different from each other, but they both use line searching, and, in both cases, information from successive line searches is not thrown away (as it is for a steepest descents minimisation) but is stored and used to shape the rest of the minimisation. This is a good idea, because it can stop the algorithm tacking back and forth, as can happen with steepest descents. Of course, each step requires more thought, because the algorithm must consider not just where it is but also how it got there. In practice, this extra thought is a good thing, and saves time in the long run. There are many different variants of this method, depending on exactly how the gradients are used. The Polak–Ribière conjugate gradient algorithm is commonly used.

BFGS (Broyden–Fletcher–Goldfarb–Shanno) or DFP (Davidon–Fletcher–Powell) are variable metric methods which are very widely used, and have been shown to be very effective. They were developed before conjugate gradient methods, and are probably more popular, because most people will adopt one minimisation method and not change it without some pressure. There is no firm evidence as to whether conjugate gradient methods or variable metric methods are better for minimising molecular structures. It is certain that both approaches are better than steepest descents. More detailed discussion of the merits of the methods are available (for example, Press *et al.*, 1992).

3.2.3 How Do We Know If We Have Arrived?

A gradient of zero means a stationary point. But there is always some uncertainty in a numerical method, so a low gradient is usually chosen to be the cut-off. This can lead to problems. Has the energy really converged? The diagram of the minimisation of cyclohexane (Figure 3.2) spent most of its time considering this question.

One way to be sure that a minimum has been reached is to perform a normal mode analysis. This distinguishes between minima, saddle points and maxima. It is not possible for this to miss anything, but it is also rather a complex and time consuming calculation, which is rarely done for a simple minimisation, provided the result looks reasonable.

3.3 WHAT CAN BE LEARNED FROM MINIMISATION?

Minimisation is the process of making small adjustments to a molecule's geometry. The process stops when all small adjustments lead to an increase in energy. Minimisation is a straightforward process, and is the limit of many molecular modelling programs, especially those written for the PC or Macintosh. It can provide very useful information.

Simply 'tidying up' a molecule is an important process which it is hard to define in a rigorous way. For two-dimensional molecular structure diagrams, 'tidying up' is not a trivial process, but for three dimensions it is relatively easy to do, simply by minimising the E_{MM} of a molecule.

If a molecule is a rigid one, then there may only be one minimum in the potential energy surface, and so minimisation is all that is required to get an idea of its properties. If it is flexible, the conformation searching techniques of Chapter 4 must be used in conjunction with minimisation.

Minimising a molecule to a form that 'looks reasonable' can be very misleading, because there may be another structure which is much more important, separated from the one found by an energy barrier.

3.4 KEY POINTS

- Minimisation of E_{MM} is a straightforward process which can be used to generate reasonable structures for molecules.
- For rigid molecules, this is all that is needed. Most molecules, however, have a measure of flexibility, and so some form of conformation searching is essential to obtain meaningful results.
- Minimisation gives energies (E_{MM}) which can be used to estimate equilibrium ratios. This is most precise for similar molecules (stereo-isomers), but can also be used with caution for other isomers.
- Differences in energies are usually more reliable than absolute energies, because some of the errors will cancel. If possible, force field investigations should be based on finding energy differences.
- Differences of differences (for example, when investigating a trend in a series of similar compounds) are even better.
- Molecular mechanics calculations should never be separated from experiments. Comparisons with experimental data in similar systems are an essential part of any molecular mechanics study.

QUESTIONS

If you have access to a molecular modelling system, try to calculate answers to these questions before looking on to the discussion of each question in the next section. If you do not have access to any programs which allow you to calculate the molecular mechanics energy, E_{MM}, of molecules, it may well be worthwhile thinking about the questions. Do the answers which molecular mechanics produces fit your expectations and prejudices?

Q3.1 Unbranched alkanes

What is the preferred shape of an unbranched alkane? Will the same geometry be preferred for alkanes of every length? What is the energetic cost of twisting the chain (Figure 3.5)?

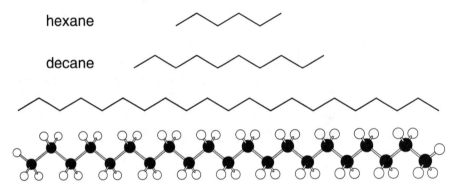

Figure 3.5 *Various alkanes*

Q3.2 Cyclohexane

Build a cyclohexane molecule in its chair conformation. Minimise the structure using the MM2 force field. Make a note of the energy, then repeat the procedure for the boat conformation of cyclohexane. Are the relative energies as you would expect? Measure the bond lengths and angles of the two molecules. How do they differ? Are there any other conformations? View the molecule as a CPK model or a ball and stick model. Is there a hole in the middle of cyclohexane (Figure 3.6)?

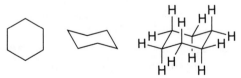

Figure 3.6 *Cyclohexane*

Q3.3 Axial and equatorial substituted cyclohexanes

Find the relative energies of equatorial and axial methylcyclohexane. Which is the preferred conformation? How does this differ for phenyl-cyclohexane, *iso*-propylcyclohexane and *tert*-butylcyclohexane (Figure 3.7)?

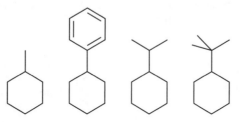

Figure 3.7 *Substituted cyclohexanes*

Q3.4 What shape is decalin?

What shape is decalin? Is there more than one accessible conformation for each diastereoisomer (Figure 3.8)?

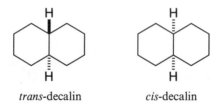

trans-decalin   *cis*-decalin

Figure 3.8 *Decalin*

Q3.5 Axial substituted cyclohexane

Can you design an alkyl sidechain for a monosubsituted cyclohexane which prefers the axial over the equatorial position?

Q3.6 Chlorocyclohexane

Do *trans-para*-dichlorocyclohexane and *para*-chlorocyclohexanone prefer to have their chlorines in equatorial or axial positions? Can you suggest a reason for the preference?

Q3.7 Cyclohexene

What shape is cyclohexene (Figure 3.9)? What orientation will substituents take up?

Figure 3.9 *Cyclohexene*

What effect would a *tert*-butyl sidechain have on electrophilic attack? Does the conformation of the ring give an explanation for the selectivity in this reaction (Figure 3.10)?

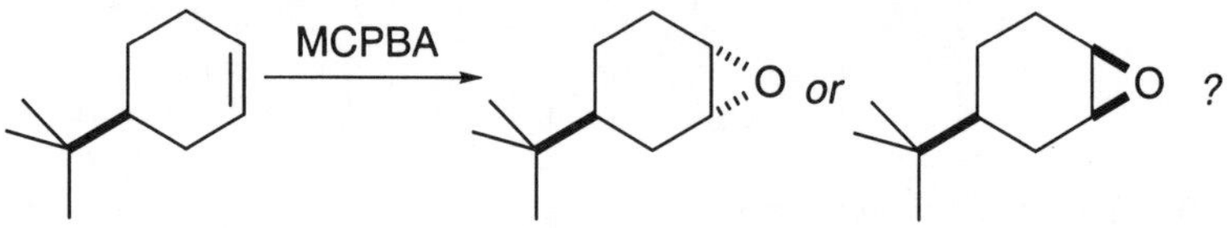

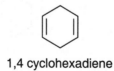

Figure 3.10 *Epoxidation of cyclohexene*

Q3.8 What shape is 1,4-cyclohexadiene?

What shape is 1,4-cyclohexadiene (Figure 3.11)?

1,4 cyclohexadiene

Figure 3.11 *1,4-Cyclohexadiene*

Q3.9 Biphenyl

What shape is biphenyl? Could a substituted biphenyl be chiral? What about bisnaphthols (Figure 3.12)?

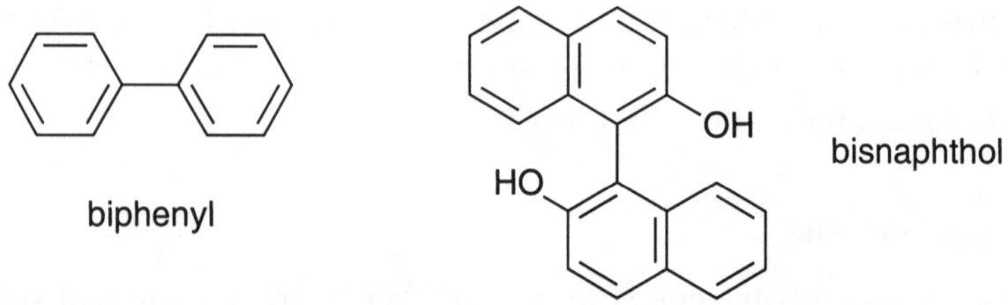

OH

bisnaphthol

HO

biphenyl

Figure 3.12

Q3.10 Annelated benzenes

Why are annelated benzenes chiral? How many benzene rings must be annelated before the molecules become chiral (Figure 3.13)?

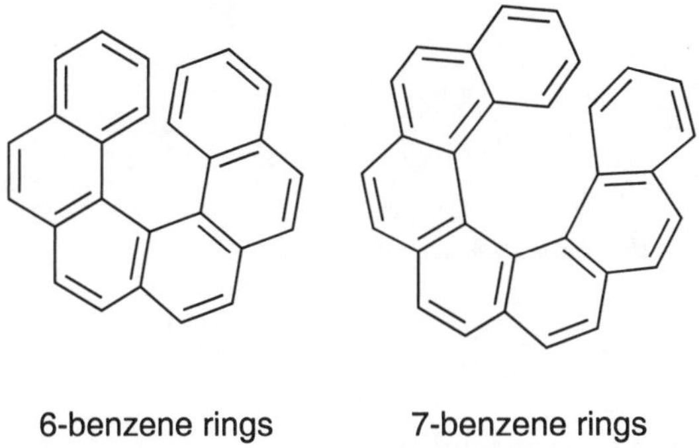

6-benzene rings   7-benzene rings

Figure 3.13 *Annelated benzenes*

Q3.11 Aromatic stacking

Construct two benzene rings close to each other and minimise them. Do they tend to stick together? What is the shape of the complex? What about naphthalene and larger fused aromatic systems?

Q3.12 Morphine

What shape is morphine (Figure 3.14)?

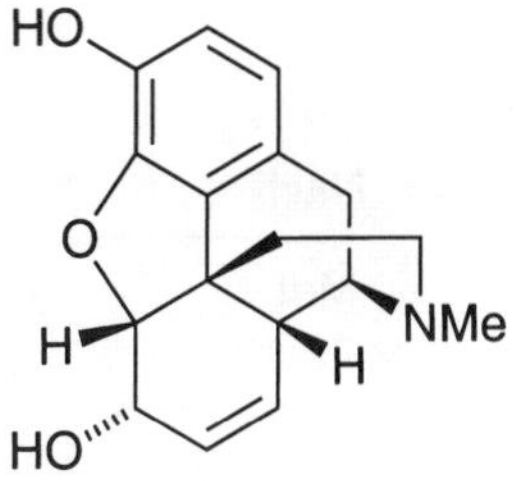

Figure 3.14 *Morphine*

Q3.13 Bredt's Rule

Bredt's Rule (Bredt, 1924) states that double bonds will not form across bridgehead carbons.

Build each of the unsaturated ring structures and record the energy of each after minimisation (Figure 3.15). Do the results account for Bredt's Rule?

Figure 3.15 *What is the elimination selectivity?*

Q3.14 An unusual rearrangement

The keto-alcohol, structure **A** in Figure 3.16, can be methylated to give the methyl ether **B**, as expected. Surprisingly, the keto-alcohol **C** reacts to give the same product. The rearrangement can be rationalised as a retro-aldol process preceding the methylation. But what is the driving force for the rearrangement? Calculate the energies of the products, or of suitable intermediates, to find out.

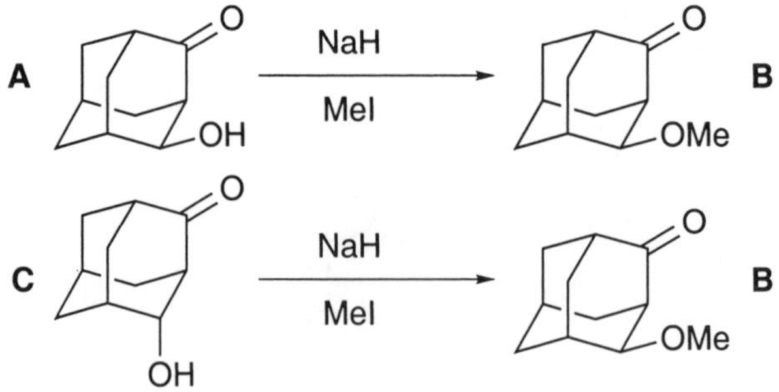

Figure 3.16

Q3.15 DNA and RNA

Build a DNA or RNA base pair and minimise them. Does this give you the structure you would expect? If sequences of base pairs are linked together, do they form a helix (Figure 3.17)?

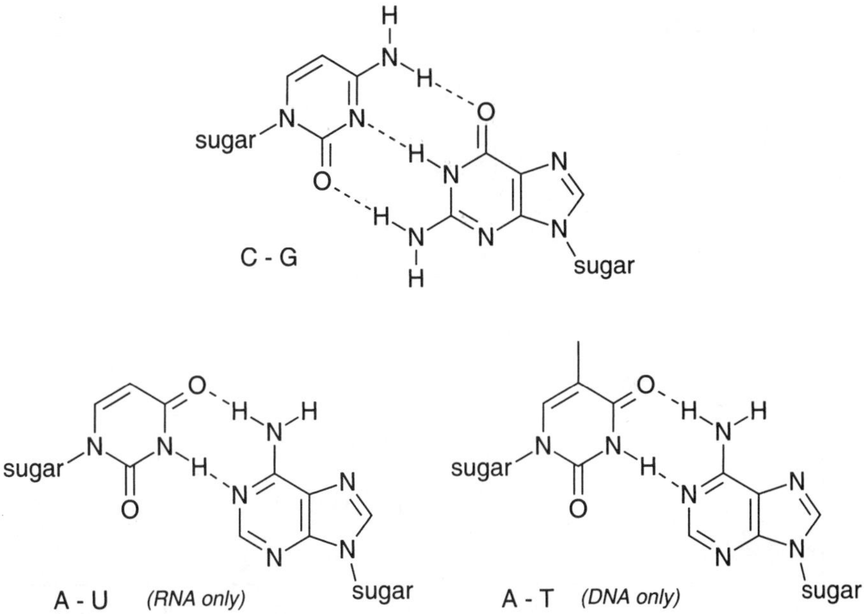

Figure 3.17 *Nucleic acid bases*

Q3.16 Protein structures

Databases of enzyme structures are available. The Brookhaven Protein Database is widely used and freely available. Do these crystal structures correspond to molecular mechanics calculated structures? Obtain a crystal structure and try to minimise it.

DISCUSSION

The calculations described in these discussions were all performed with MacroModel (Mohamedi *et al.*, 1990), using the MM2* force field, unless indicated otherwise. Other molecular modelling packages and force fields should give comparable results.

D3.1 Unbranched alkanes

What is the preferred shape of an unbranched alkane? Will the same geometry be preferred for alkanes of every length? What is the energetic cost of twisting the chain?

Unbranched alkanes prefer extended conformations, as illustrated in the question, because this is the conformation which minimises all adverse steric interactions. Rotation around a central bond gives an

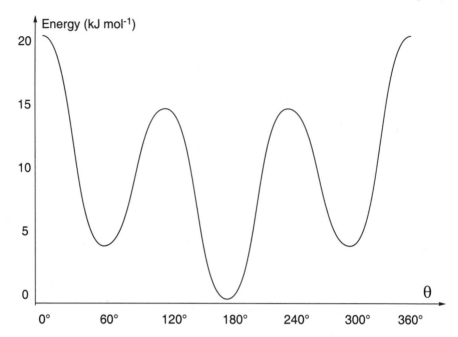

Figure 3.18

energy profile resembling Figure 3.18, which makes it clear that the optimal torsion angle for a C—C—C—C fragment is 180°, although the penalty for twisting through 120° to 60° or 300° is not too great. This suggests that the extended conformation should always be the global minimum for unbranched alkanes.

However, it is easy to forget that there is an *attractive* van der Waals interaction as well as a repulsive one. If the chains of the alkane are long enough, the attractive interaction can more than counterbalance the energetic cost of a few twists in the chain. A chain can be reversed with two twists, but this gives a rather strained structure (*syn*-pentane interaction, diaxial across a cyclohexane). Four twists do not cost very much energy (about $14\,\mathrm{kJ\,mol^{-1}}$) and neatly reverse the chains (Figure 3.19a). Each methylene pair separated by the sum of their van der Waals radii, gives a favourable interaction of about $4\,\mathrm{kJ\,mol^{-1}}$. Four methylene pairs are required to 'pay' for the unfavourable interaction of the twist, so a hairpin conformation may be expected to be lower in energy than an extended conformation for alkanes larger than about $C_{18}H_{38}$ (Figure 3.19b and Plate 1). This turns out to be the case (Goodman, 1997). The precise length of the chain for which the folded structure is preferred depends on the force field, and is a rather sensitive test of the properties of a force field. Longer alkanes will prefer more complex folded structures, and this is discussed further in the Section 4.15.

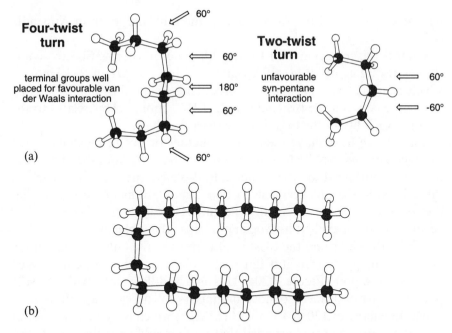

Figure 3.19 *The lowest energy conformation (global minimum) for* $C_{18}H_{38}$

D3.2 Cyclohexane: comparison of chair and boat forms

Minimisation of cyclohexane can give a chair or a twist boat. A non-twist boat is not normally formed, as it is a saddle point on the potential energy surface not a minimum. Table 3.1 gives the energy differences and the chair to twist-boat ratio calculated at 300 K for a number of force fields. The approximate ratio can be read from graph in Appendix A.3.

Table 3.1 *Results of some force field calculations*

Force field	chair (kJ mol^{-1})	twist-boat (kJ mol^{-1})	Energy difference (kJ mol^{-1})	chair : twist-boat ratio at 300 K
MM2*	27.41	49.82	**22.41**	8 000 : 1
MM3*	33.66	57.75	**24.09**	16 000 : 1
AMBER*	9.49	33.71	**24.22**	16 000 : 1
AMBER* (no hydrogens)	22.84	42.46	**19.62**	2 600 : 1
MMFF	− 14.90	9.91	**24.81**	21 000 : 1
COSMIC	6.09	30.44	**24.35**	21 000 : 1
Chem3D	27.43	49.85	**22.41**	8 000 : 1
Sybyl	5.94	38.16	**32.22**	400 000 : 1

References for all of the force fields in the table are given in Appendix A.7.

What is the correct answer for the difference in energy? Spectroscopic studies (Squillacote *et al.*, 1975) give a value for ΔH of $23\,kJ\,mol^{-1}$, which is similar to many of the force fields listed above. The entropy difference between the two states is $22\,J\,K^{-1}\,mol^{-1}$. This relates to the greater flexibility of the twist boat than of the chair.

Some force fields have the option whether or not to include explicit hydrogen atoms. AMBER is an example of this. Without the hydrogens only six atoms need to be considered in this case, instead of eighteen, so the calculation is quicker. The carbon atoms are made larger, so that they occupy the space of a methylene unit as one atom. It is faster but less accurate than an explicit hydrogen model, as the table illustrates.

Whilst the chair and the twist boat forms are the only minima on the potential energy surface, it is fairly easy to create other structures, such as a flat hexagon (Figure 3.20). This is a saddle point on the MM2* potential energy surface, so most minimisers will accept it as a minimum. It has an energy of $130.40\,kJ\,mol^{-1}$ (MM2*) which is a clue that it is not an important structure, provided that you already have the energy of a chair or a boat with which to compare it.

Comparing Bond Angles and Lengths. The bond angles and lengths are almost exactly the same for the chair, twist-boat and the flat saddle point. If a structure is drawn with an umbrella-shaped carbon atom, some molecular mechanics programs will minimise the energy while maintaining the umbrella-shape. This will give a very high energy. It is always important to check that the minimised structure looks reasonable, and does not have any such unrealistic features.

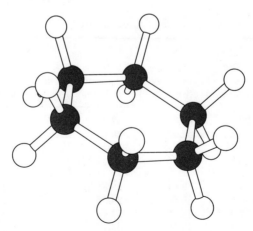

Figure 3.20 *A high-energy saddle point on the cyclohexane potential energy surface*

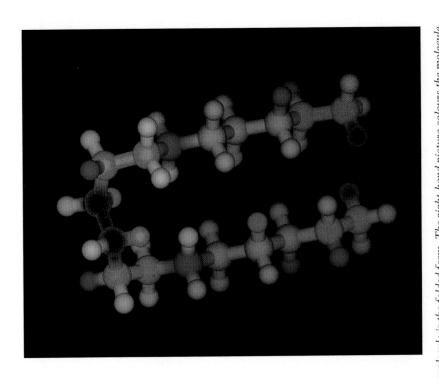

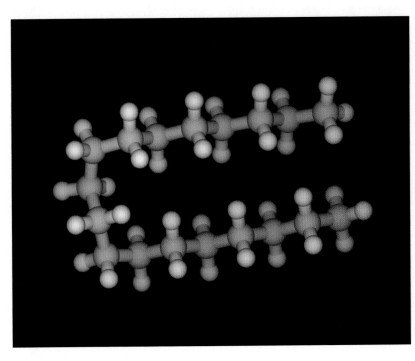

Plate 1 Octadecane ($C_{18}H_{38}$). The global minimum conformation for this molecule according to the energy of each part. Grey is unstrained, yellow and red are high energy parts, blue shows a favourable interaction (See Chapter 3, Q3.1 and discussion D3.1, and Chapter 4, Q4.1 and discussion D4.1)

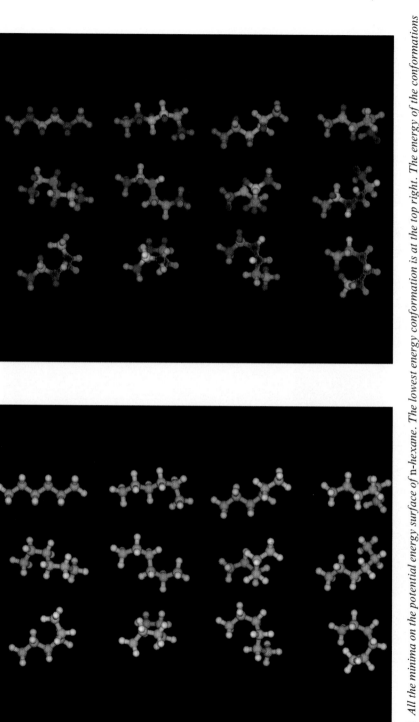

Plate 2 *All the minima on the potential energy surface of n-hexane. The lowest energy conformation is at the top right. The energy of the conformations increases down the columns. The highest energy conformation is at the bottom left. The colours of the second set of structures indicate the strain in each conformation (See Chapter 4, Q4.1, and discussion D4.1)*

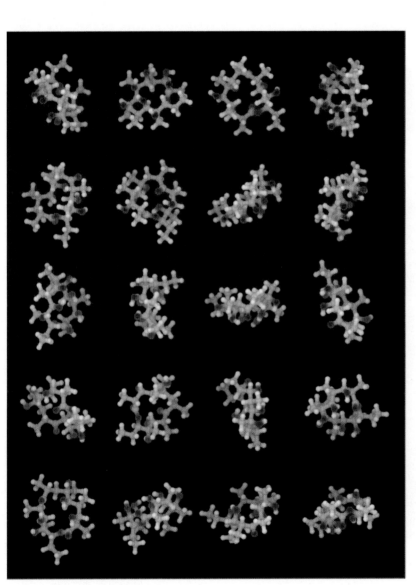

Plate 3 *The 20 lowest energy conformations for erythronolide (Chapter 4, Q4.7). These are only a few of the conformations which are significantly populated (See Chapter 4, Q4.7, and discussion D4.7)*

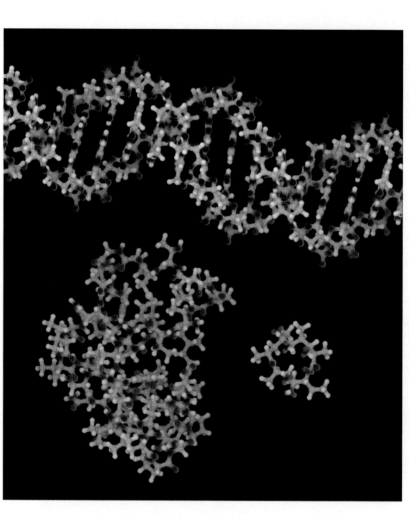

Plate 4 *Comparison of the sizes of erythronolide (Chapter 4, Q4.7), crambin (Chapter 3, Q3.16) and DNA (Chapter 3, Q3.15) (See Chapter 4, Q4.13, and discussion D4.13)*

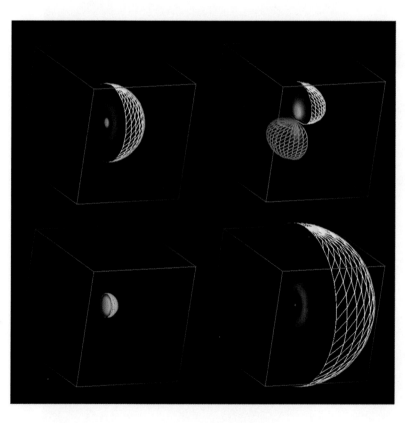

Plate 5 1s, 2s, 3s and 2p atomic orbitals, all drawn to the same scale. The lattice surface encloses 95% of the electron density of the orbital. The yellow colour in the cross-section shows the electron density. The 3s orbital appears to have less electron density, because the same amount is spread out over a much greater volume

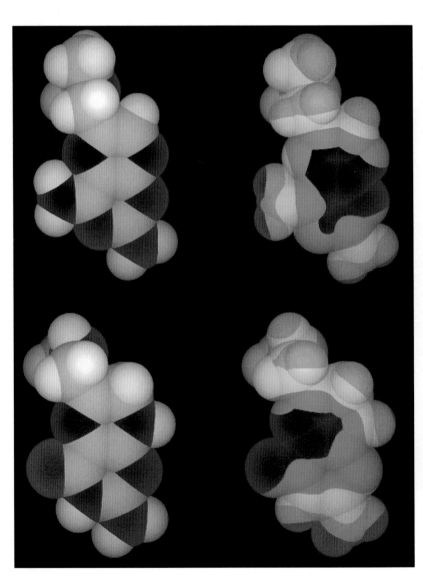

Plate 6 *Methotrexate and dihydrofolate. The two compounds appear rather similar when shown as atom types (above), but one appears to be upside down when the electrostatics are compared (See Chapter 8, Section 8.4.2)*

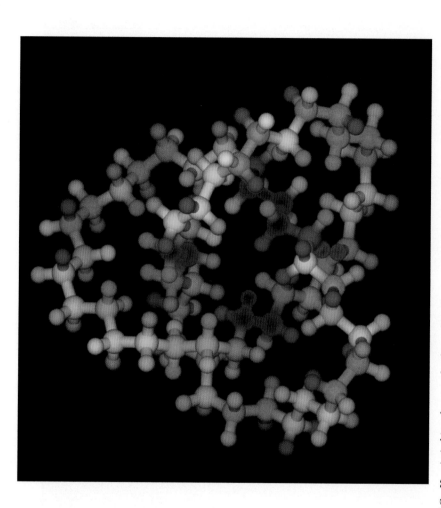

Plate 7 *Cyclohexacontane* ($C_{60}H_{120}$) *tied in a knot and coloured by the energy in the different parts of the molecule. Another representation of this molecule appears on the front cover*

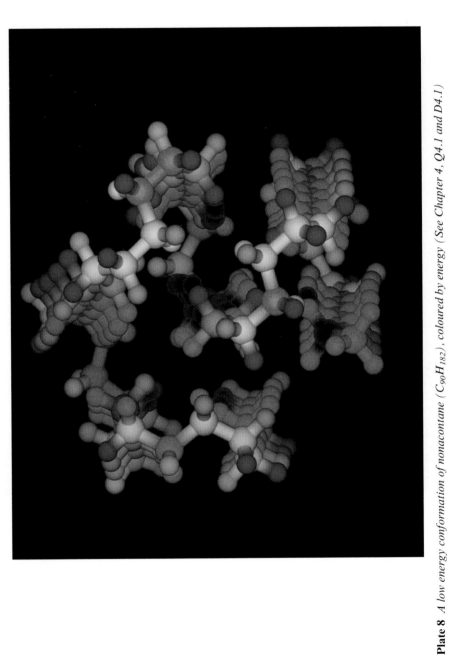

Plate 8 *A low energy conformation of nonacontane* $(C_{90}H_{182})$, *coloured by energy (See Chapter 4, Q4.1 and D4.1)*

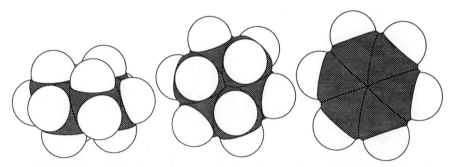

Figure 3.21 *CPK models of cyclohexane (edge-on), cyclohexane and benzene*

Display of Different Fractions of the van der Waals radius. There is not really a hole through the middle of cyclohexane, or benzene. A CPK model illustrates this (Figure 3.21). Significant electron density extends right across the centre of the ring.

D3.3 Axial and equatorial substituted cyclohexanes

Table 3.2 shows the minimised structural energies for axial and equatorial substituted cyclohexanes. The energy difference for methylcyclohexane is $7.44\,\text{kJ}\,\text{mol}^{-1}$ which corresponds to a ratio of 20:1 at room temperature (this can be read from the chart in Appendix A.3). This is considerably less than the difference between chair and twist-boat cyclohexane, calculated in D3.2. For phenylcyclohexane, the ratio is much higher, showing that a phenyl group is larger than a methyl group, by this method of estimation. There are two entries for the energy of the axial form, because minimisation can give two different structures (Figure 3.22)

It would be easy to be satisfied with the first one of these that is found, and not to bother investigating whether more conformations are possible. In this particular example, the difference is not too crucial, because both forms are high in energy. There are many situations where this

Table 3.2 *Structural energies for substituted cyclohexanes* $(\text{kJ}\,\text{mol}^{-1})$

R	(axial)	(equatorial)	Energy difference
Me	36.27	28.83	7.44
Ph	59.86, 63.19	43.40	16.46
^{i}Pr	52.68, 65.72	45.57, 45.48	7.20
^{t}Bu	76.60	55.70	20.90

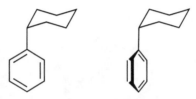

Figure 3.22 *Two conformations of axial phenylcyclohexane*

might be crucial. This is discussed in greater detail in the next chapter. Multiple conformations are also important for *iso*-propyl-cyclohexane, and these are illustrated in Figure 3.23.

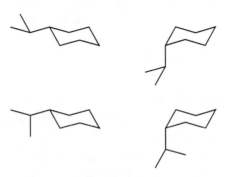

Figure 3.23 *Conformations of* iso-*propylcyclohexane*

Is the *iso*-propyl group larger than a methyl group? The obvious answer is 'yes' but the energy difference in the table suggests that it is rather similar to a methyl group. The reason for this is, in part, that the *iso*-propyl group is only as large as a methyl group when pointing in the right direction. This is not the whole explanation, because the energy difference is less for the *iso*-propyl group than for the methyl group. Another contributing factor is that the equatorial *iso*-propyl group suffers from unfavourable steric interactions which the equatorial methyl group escapes.

D3.4 What shape is decalin?

In Chapter 1, Barton's calculation of the decalin structure was described (Barton, 1948). The calculations were very hard to carry out, as they were performed in the days before cheap digital computers.

It is now straightforward to calculate the relative energies of the different forms of decalin (Figure 3.24). The difference in energy between

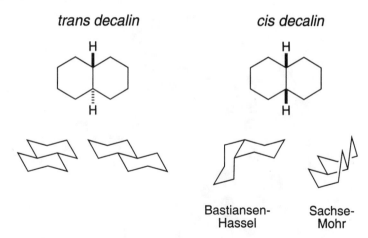

Figure 3.24 *Decalin*

cis- and *trans-*decalin is calculated using MM2* to be $11.5\,\mathrm{kJ\,mol^{-1}}$, with the same preference as the experimental difference of $21\,\mathrm{kJ\,mol^{-1}}$ measured by comparison of the heats of combustion of the two compounds. Neither *trans-* nor *cis-*decalin is chiral, but the conformation of the *cis* form is chiral. The two different forms can interconvert by flipping both rings.

D3.5 Axial substituted cyclohexane

Is it possible to design an alkyl sidechain for cyclohexane which prefers the axial over the equatorial position? A rigid V-shaped side chain may prefer the axial position. Such a sidechain has been designed by Biali, based on adamantane units, and is calculated to prefer the axial position (Biali, 1992). Figure 3.25 shows hydrogens only on the cyclohexane ring, for clarity.

D3.6 Chlorocyclohexane

Do *trans-para-*dichlorocyclohexane and *para-*chlorocyclohexanone prefer to have their chlorines in equatorial or axial positions?

Both these compounds prefer conformations with axial chlorines (Figure 3.26), contrary to a simple expectation that large groups tend to prefer the equatorial position. This preference has been shown experimentally (Abraham and Rossetti, 1973). It may be attributed to an attraction between the chlorine's negative partial charge and the positive partial charge of the carbons attached to electronegative groups.

Figure 3.25 *A cyclohexane substituent which prefers to be axial*

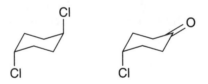

Figure 3.26

D3.7 Cyclohexene

Cyclohexene is often represented by looking along the plane of the double bond, and this gives a structure very similar to the calculated one (Figure 3.27).

Tertiary butyl substitution locks the ring, but the group is so far away that it does not appear to block attack at either face of the double bond (see, for example, Klunder *et al.*, 1985 and Figure 3.28).

D3.8 What shape is 1,4-cyclohexadiene?

1,4-Cyclohexadiene is almost completely flat, even though it is often drawn in a boat shape. Evidence for this comes not only from molecular mechanics calculations, but also from high level *ab initio* molecular orbital theory calculations and from X-ray crystal data (Figure 3.29).

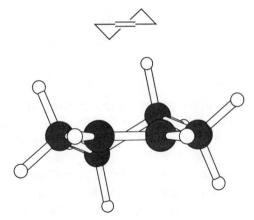

Figure 3.27 *The conformation of cyclohexene*

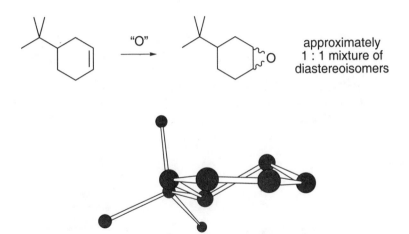

approximately
1 : 1 mixture of
diastereoisomers

Figure 3.28 *Electrophilic attack on cyclohexene*

incorrect   correct

Figure 3.29 *1,4-cyclohexadiene*

D3.9 Biphenyl

Both biphenyl and bisnaphthol are twisted around the central carbon–carbon bond, despite the advantage of conjugation that could be gained from being flat. The torsion angle between the aromatic rings of biphenyl is 38°, according to MM2*, and for bisnaphthol it is 85°. This twist means that substituted biphenyls and bisnaphthols can be chiral.

D3.10 Annelated benzenes

Four fused aromatic rings have an helical conformation, but only three are flat. The barrier to interconversion for four fused rings can be estimated as $11\,\mathrm{kJ\,mol^{-1}}$ by minimising the structure whilst constraining it to be flat. This suggests that it will invert something like 10^{11} times a second at room temperature (Appendix A.4), and so it will not be easy to measure the effect of the chirality. Five fused rings have a much higher barrier for racemisation and the enantiomers can be isolated (Figures 3.30 and 3.31).

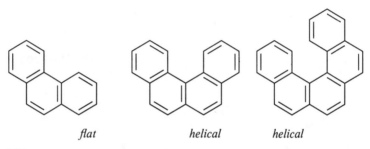

flat   *helical*   *helical*

Figure 3.30

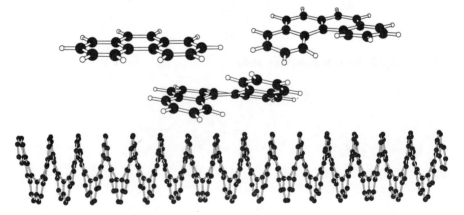

Figure 3.31 *Conformations of annelated benzenes*

D3.11 Aromatic stacking

Construct two benzene rings close to each other and minimise them. Do they tend to stick together? What is the shape of the complex?

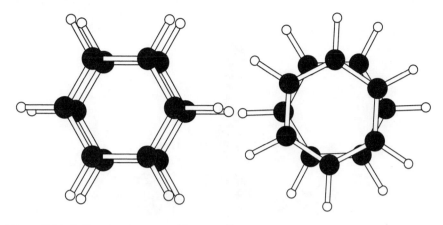

Figure 3.32 *Eclipsed and staggered benzene dimers*

MM2* and many other force fields will create structures like Figure 3.32 for a van der Waals dimer of benzene. The favourable van der Waals interaction is maximised by stacking the rings neatly. The two forms have very similar energies, and the potential energy surface is very flat in this region, making it hard to decide when minimisation should be terminated.

However, benzene rings do not stack up like the two calculated minima, but prefer to be offset or perpendicular. The same is true for porphyrins (Hunter and Sanders, 1990). The energy differences are small, but are very significant for the stacking of base pairs in DNA and the interactions of aromatic rings in proteins. This is a major deficiency of most force fields.

D3.12 Morphine

Morphine looks very unlike its conventional representation.

The left hand structure in Figure 3.33 is a conventional way of drawing morphine, although it is equivalent to the other three. The right hand structure gives an impression of the T-shape of the molecule. The shape

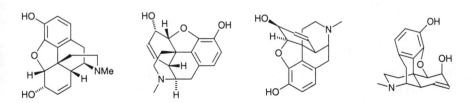

Figure 3.33 *Representations of morphine*

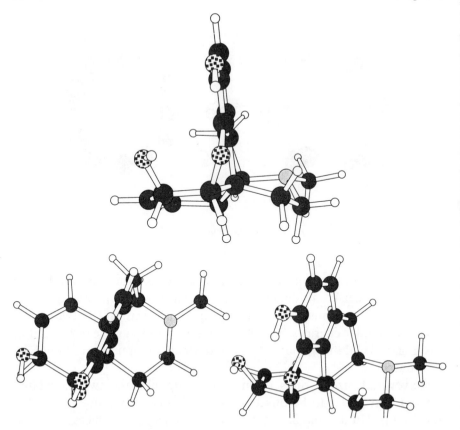

Figure 3.34 *Three-dimensional representations of morphine*

of the molecule must be the key to its activity, but is poorly expressed by the two-dimensional picture. Three-dimensional pictures (Figure 3.34) illustrate this much more clearly: Morphine was simply sketched into a computer and minimised to obtain these structures. Complicated poly-cyclic structures like this are particularly prone to developing umbrella-shaped carbons atoms, or intersecting rings. These features would give the structure a very high energy, but this might not be noticed, as there is nothing with which to compare it. Conformations should always be checked to be sure that they are reasonable.

D3.13 Bredt's Rule

Bredt's Rule states that double bonds will not form across bridgehead carbons (Bredt, 1924).

Using the MM2* force field, structure **A** in Figure 3.35 has an energy of $106.62 \, \text{kJ} \, \text{mol}^{-1}$ and **B** an energy of $216.31 \, \text{kJ} \, \text{mol}^{-1}$. If the two

Figure 3.35

species are in their thermodynamic ratio, then **A** should be favoured by a factor of about 10^{20}, so a mole of the mixture would contain less than 10 000 molecules of **B**.

This huge energy difference clearly shows a strong preference for **A**. However, the ratio is probably not given very accurately. Force fields are parameterised for ground states, for which it is easiest to get experimental data. If a double bond is twisted, then its energy goes up a lot, but it is not clear exactly how high. This result should be believed, insofar that it suggests that the bridgehead double bond is much higher in energy. The force field probably could not give any precise idea about the relative energies of different bridgehead double bonds, or other high-energy species.

D3.14 An unusual rearrangement

The keto-alcohol, structure **A** in Figure 3.36, can be methylated to give the methyl ether **B**, as expected. Surprisingly, the keto-alcohol **C** reacts to give the same product (Duddeck *et al.*, 1979).

The molecule could rearrange through a retro-aldol process (Figure 3.37). But what is the driving force? The way to start is to calculate the relative energies of **B** and **D**, and this suggests that **B** is favoured, but only by 2.1 kJ mol^{-1}. This corresponds to a ratio of 5 : 2 in favour of the observed product. Whilst no **D** was observed in the reaction mixture, the

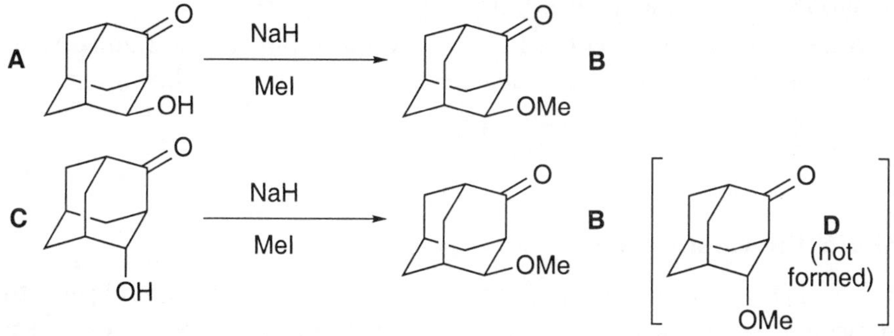

Figure 3.36

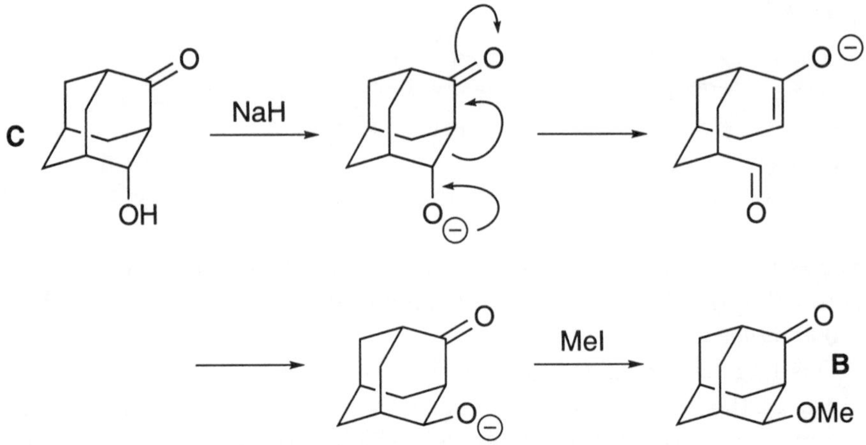

Figure 3.37 *Mechanism for the rearrangement*

yield of **C** was less than 100% and so this may be enough to account for the experimental result. Duddeck also reports that **A** and **C** could be equilibrated directly, and the thermodynamic ratio favoured **A** in ratios varying from 3:1 to 7:1, which suggests that there is not an overwhelming thermodynamic preference for the equatorial form over the axial form.

D3.15 DNA and RNA

A single DNA base pair may well minimise to remain in the Watson and Crick form (Watson and Crick, 1953). However, this is probably not its form in solution, since a normal gas-phase minimisation will overemphasise the strength of the hydrogen bonds which hold the base pairs together. If a solvent model is introduced, then the hydrogen bonds will be weakened, and a surface area term may be introduced to simulate hydrophobicity. This will tend to mean that the bases will prefer to stack, rather than form a hydrogen-bonded base pair.

A sequence of base pairs can easily be built, but they are unlikely to minimise into a helix, unless they were constructed with this in mind. A very wide range of conformations would be open to them. This question is addressed in the following chapters.

D3.16 Protein structures

It has been shown that crystal structures correspond quite closely to molecular mechanics minima. This is not too surprising, because protein structures are often obtained by fitting a structure to X-ray data using a force field. However, many proteins contain metals and these cannot be modelled well by most molecular mechanics methods.

The minimisation of a protein may be very slow, because a small movement of a residue in the centre of a protein can lead to a large change in energy, if bad steric effects are introduced. Force fields often use electrostatic cut-offs, to reduce the time required for energy calculations. This may have a disastrous effect on the minimisation of a protein, because the cut-off may fall inside the structure, and so a small change can lead to a discontinuity in the potential energy surface. Cut-offs should always be set to include the whole structure, unless this problem has been specifically addressed in the design of the force field.

For example, crambin (Figure 3.38) is a small protein for which a high quality crystal structure is available (Teeter, 1984). Minimisation of this molecule using MM2* and the default van der Waals and electrostatic cut-offs for small molecules gives an energy of $-1614.9\,\mathrm{kJ\,mol^{-1}}$ and a geometry which is very similar to the crystal structure. If this structure is reminimised in exactly the same way, an energy of $-1620.8\,\mathrm{kJ\,mol^{-1}}$ is obtained. If this new structure is reminimised, the energy returns to $-1614.9\,\mathrm{kJ\,mol^{-1}}$! This result arises because the cut-offs fall within the

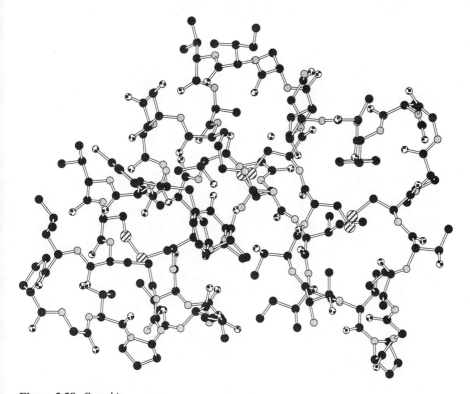

Figure 3.38 *Crambin*

protein, which is too large for the standard distances for these cut-offs. The parameters may be adjusted, so that the cut-offs include the whole molecule. If this is done, a consistent energy of $-1707.4 \, kJ \, mol^{-1}$ is obtained for the structure. The geometry of the structure barely changes during these repeated minimisations.

Assuming that a crystal structure does not change significantly on minimisation, it is possible to perform site-directed mutagenesis. An amino acid can be edited so that it becomes a different residue, and the structure can be reminimised. This gives an idea of the structure of the mutant. It does not answer an important question: 'Will the mutant have the same tertiary structure as the wild type protein?'

CHAPTER 4

Conformation Searching

4.1 INTRODUCTION

Most molecules of interest to chemists, biochemists and pharmacologists are not described very well by a single low energy conformation. It would be useful to know which conformation has the lowest energy of all (the global minimum), and also to have a list of all the structures which are close in energy to the global minimum. This can sometimes be achieved by conformation searching. Many algorithms are available that generate structures of low energy, but the problem is a difficult one, and is often the rate-limiting step in a molecular mechanics study.

Conformation searching is a crucial step in most molecular mechanics studies. Without a full conformation search, there can be no reason to expect that the conformations of a molecule which are being considered represent the lowest energy, and, therefore, the most important ones. The examples in the previous chapter include some examples of this. A flat cyclohexane ring, or one with an umbrella-shaped carbon, will be very high in energy, but this can only be appreciated if there is a lower energy structure with which to compare them. In this case, it should be obvious to an experienced chemist that there is something wrong with the high energy structure, which should prompt further exploration. The substituted cyclohexanes (Q3.3) are much more subtle. It is not immediately obvious that either of the low energy conformations of equatorial and axial *iso*-propylcyclohexane do not represent the only low energy structure. With more complex structures, it is not possible to tell by inspection whether a conformation is likely to be the lowest in energy of all possibilities (*i.e.* the global minimum). Automated methods of searching conformation space are essential to find all the low energy conformations, and may often produce surprises.

The following procedures are described in this chapter:

- Systematic Search: plod through all possibilities—reliable, but slow.
- Monte Carlo Search: make use of information gathered as the search proceeds—much faster, and often the method of choice.

- Molecular Dynamics (described in detail in Chapter 5): not an ideal method for conformation searching, unless many experimental data (*e.g.* diffraction patterns or nOe data) are available.
- Genetic Algorithms.
- Distance Geometry.
- Rule-based Methods: very rapid method of generating a reasonable structure, but not necessarily the global minimum.

4.2 WHY DO A CONFORMATION SEARCH?

In the last chapter, minimisation was compared to walking around a hillside in a thick fog, trying to find the way down hill. This is not a trivial problem. In this chapter (*Further adventures of a molecular-modelling hill walker*), there is a much more challenging task. It is necessary to go down hill, but there are many valleys which do not interconnect. The task is to find the lowest point of all, where a warm bath and hot meal are waiting. Just walking downhill will find the closest valley, and the lowest point of this valley. This does not guarantee finding the final destination, which may be over the next hill.

The lowest energy point on a potential energy surface is called the *global minimum*. Each potential energy surface can only have one. However, there may be many *local minima*. These are minimum energy points on the potential energy surface which are higher in energy than the global minimum (Figure 4.1).

To take a simple example, the global minimum form of methylcyclohexane is the chair conformation with the methyl group equatorial. The chair conformation with an axial methyl group and the twist boat conformation are both local minima (Figure 4.2).

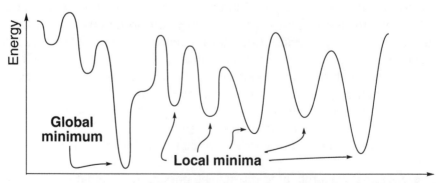

Figure 4.1 *A potential energy surface*

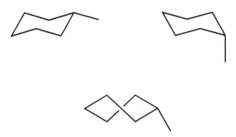

Figure 4.2 *Some conformations of methylcyclohexane*

What will the ratio of these forms be? This can be calculated from the energies of the minima using Boltzmann factors (Equation 4.1):

$$\text{Boltzmann Factor} = \exp\left(\frac{-E}{RT}\right) \qquad (4.1)$$

where R is the gas constant, $8.314\,\mathrm{J\,K^{-1}\,mol^{-1}}$. At room temperature, $T \approx 300\ \mathrm{K}$.

The energies of the different forms were calculated using the MM2* force field and the MacroModel program (Figure 4.3). At room temperature, $RT = 2.49\,\mathrm{kJ\,mol^{-1}}$. The ratio of these three forms will be the same as the ratio of their Boltzmann factors (Table 4.1).

Even at room temperature, the dominant conformation is clearly the chair with the equatorial methyl group. Have all the possible conformations been considered? In fact, there are many more minimum energy conformations which are accessible to methylcyclohexane. This question is examined in detail in the discussion section (D4.2).

The shape of molecules is a standard part of undergraduate examinations, although it is usual to provide questions which can be answered by thinking, drawing or building models. It may become possible to set questions which require much more demanding conformation analysis, if the standard equipment for an examination includes a workstation. For example, a key step in one of Still's syntheses (Still and Romero, 1986) is illustrated in Figure 4.4.

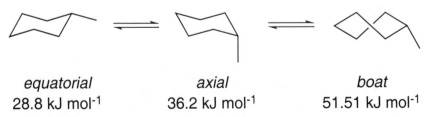

equatorial	axial	boat
28.8 kJ mol⁻¹	36.2 kJ mol⁻¹	51.51 kJ mol⁻¹

Figure 4.3 *Relative energies of methylcyclohexane conformations*

Table 4.1 *Ratio of forms of methylcyclohexane*

Conformation	Boltzmann factor	Ratio relative to boat
boat	$\exp\left(\dfrac{51.5}{2.49}\right)$	1
axial	$\exp\left(\dfrac{36.2}{2.49}\right)$	470
equatorial	$\exp\left(\dfrac{28.8}{2.49}\right)$	9100

MCPBA epoxidises the outer face of each double bond, forming a three-membered ring stereospecifically. A calculation of the conformation of the starting material reveals which face of each double bond is most accessible, and so can be used to predict, or rationalise, the stereochemistry of the product. An examination question could be based on this. A much harder question could also be asked. The natural product monensin closely resembles the final product, but not all of the stereocentres have been formed with the correct configuration. Is it possible to design a related precursor which would give the correct stereochemistry for monensin? In order to answer this sort of question, it is necessary to be able to find the global minimum and also all the low energy local minima for such large and flexible molecules.

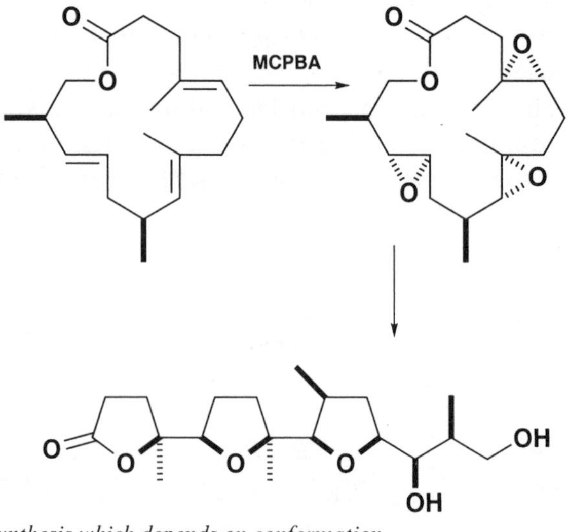

Figure 4.4 *A synthesis which depends on conformation*

4.3 SYSTEMATIC SEARCHING

Conformations will differ principally in their torsion angles. This is obvious for cyclohexane and simple organic compounds. It is not obvious for other classes of compounds. However, it has been shown that interest in 'bond-stretch isomers' is not supported by the data (Yoon and Parkin, 1991). If it is true that a molecule's conformations differ only in their torsion angles, a strategy for finding all the minima of a particular structure becomes clear: generate conformations with all possible combinations of torsion angles and minimise them all. This should generate all possible minima, including the global minimum. There are an infinite number of possibilities for combinations of torsion angles, unless torsion angles are restricted to discrete values, for example, multiples of 60°. This would mean that six different conformations must be tried for each flexible torsion angle, and so the difficulty of performing the conformation search is approximately:

$$\text{Difficulty} \approx 6^N \tag{4.2}$$

where N is the number of flexible torsion angles.

This means that a very large number of structures must be minimised to be sure of finding the global minimum, even for quite simple molecules. Some torsion angles need not be rotated, because they are attached to groups which are nearly spherical, such as methyl groups. Butane has one flexible torsion angle, and so a full conformation search is fairly straightforward, with a difficulty of about six. Hexane has three flexible torsion angles, and the difficulty is about 216. Undecane, an eleven carbon atom hydrocarbon, has a difficulty of over a million. It is possible to make the process more efficient, by avoiding combinations of torsion angles which always lead to high energies (Lipton and Still, 1988), but this does not change the approximate expression for difficulty in Equation 4.2. The principle of systematic searching is illustrated in Figure 4.5.

4.3.1 Rings

If a molecule contains a ring, it is not possible to twist individual torsion angles without changing others. This means that the conformation space is much smaller than for an acyclic system of a similar size, and so conformation searching is rather easier. Experience suggests that the equation for difficulty may be modified to:

$$\text{Difficulty} \approx 6^{N_t - 5N_r} \tag{4.3}$$

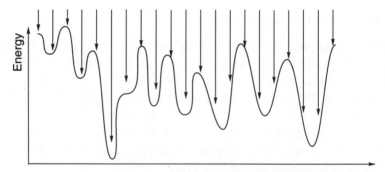

X (some description of the molecular structure)

Figure 4.5 *Systematic searching. If all the conformations indicated are generated and minimised, all the minima for this potential energy surface will be found*

where N_t is the number of flexible torsion angles and N_r is the number of flexible rings.

For example, a conformation search on each of the two esters illustrated in Figure 4.6 would take about the same length of time, even though the cyclic molecule is rather larger. The difficulty is given to six significant figures, which is somewhat misleading. The expression for difficulty is a useful rule of thumb, but should only be used as a guide to the order of magnitude of difficulty of a search. This is very useful, because it will give an idea of whether a system can be investigated in a few minutes or would require several years.

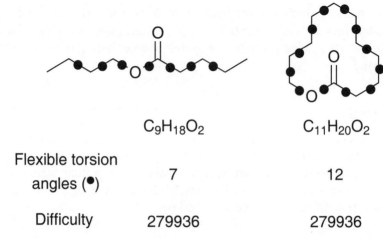

	$C_9H_{18}O_2$	$C_{11}H_{20}O_2$
Flexible torsion angles (●)	7	12
Difficulty	279936	279936

Figure 4.6

How is it possible to twist torsion angles in a molecule when there is a ring? One solution to the problem is to break the procedure up into stages (Still and Galynker, 1981). First, a bond in the ring is selected and broken. The cyclic structure is now acyclic, so all the remaining torsion angles can be twisted. The bond which was broken is now reformed. If the atoms which need to be joined are a long distance apart, then the structure is unlikely to minimise to a low energy, so many conformations can be disregarded without minimisation. If the atoms at the end of the broken bond are close, then they are rejoined, and the new structure minimised. This turns out to be an efficient strategy for searching the conformation space of cyclic compounds, which works for all ring sizes. It can also be applied to compounds containing multiple rings, simply by breaking enough bonds to reduce the structure to an acyclic form before any torsion angles are adjusted, and then reforming all of the bonds before minimisation. It is best to choose bonds to break that are away from any chiral centres, but this is not always possible. For small rings, (four-, five- or six-membered rings), the procedure works, but requires many minimisations considering the relative simplicity of the conformation space of such molecules. In these cases, alternative methods of distorting the rings may be preferable, and these are described later. The ring breaking and torsion angle twisting strategy has the great strength that it is systematic, and so is guaranteed to find all possible conformations of a molecule, provided a sufficiently small increment is chosen for the torsion angles.

4.4 MONTE CARLO SEARCHING (SEMI-RANDOM SEARCHING)

It is possible to do conformation searches more rapidly than the brute force method of systematic searching. One of the best algorithms for this is called Monte Carlo searching or semi-random searching. It is based on a key idea:

> Low energy conformations have
> structural features in common

If a reasonable, low energy structure has been found for a molecule, then it is likely that other low energy structures can be found by making a small alteration to the conformation, perhaps by moving one atom relative to the others, or by twisting just one or two torsion angles. The alterations to the structure may be chosen randomly, which gave rise to the name of this procedure, and a number of approaches have been tried (Saunders, 1987; Ferguson and Raber, 1989; Chang *et al.*, 1989). Monte

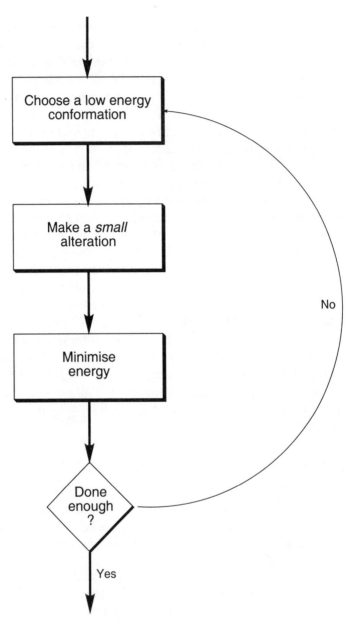

Figure 4.7 *Flow chart for Monte Carlo searching*

Carlo methods are also widely used in statistical mechanics, but the phrase has a somewhat different meaning in this context. Figure 4.7 shows a flow chart for Monte Carlo searching.

As new low energy structures are found, this information is used to help find the next low energy structure. This is in contrast to a systematic search, for which the plan of the search is decided at the beginning, and none of the information which is discovered as the search proceeds is used. It may, therefore, be expected that Monte Carlo methods should work more quickly, and this is found to be the case. The expression for difficulty given above (Equation 4.3) is still a useful rule of thumb for assessing the time required for a conformation search.

A new problem arises. With a systematic search, it is obvious when the search is complete: all possible conformations have been investigated. With a Monte Carlo search, there is no such straightforward criterion for deciding that the procedure should be stopped. After a while, the search will stop finding new low energy structures. The usual criterion used to decide when a search is complete is to examine the low energy structures and to check that all have been generated more than once.

4.4.1 Alternative Distortions

Twisting torsion angles is an efficient method of mutating molecules for a Monte Carlo search, particularly for chains and medium to large rings. The method is least effective for small rings and fused rings (although it still works), because a high proportion of the mutations will lead to structures for which the ring cannot close. In these cases, alternative strategies may be more effective

One such strategy is 'Corner Flapping' (Goto and Osawa, 1989, 1993). A cyclohexane ring may be transformed into a boat by flapping a corner of the structure (Figure 4.8). This seems to be a good way of generating reasonable ring conformations. However, it is not clear that it will necessarily generate all possible conformations.

A more sophisticated way of finding new conformations is to follow the normal modes of a ground state (Kolossváry and Guida, 1993, 1996). By calculating the normal modes of vibration of a conformation, it is possible to see the way in which it can distort for a low energetic cost. Such a movement will usually require the concerted distortion of several parts of the molecule. Distortion in these directions may lead to new conformations, and also information about the interconversion pathways of the conformations. This method appears to be particularly useful for fused ring systems.

Figure 4.8 *Corner flapping*

4.5 MUTATION METHODS

The premise behind Monte Carlo methods can be extended. If low energy conformations have features in common, then similar compounds with different structures may also have related conformations. It is often the case that it would be interesting to perform a conformational analysis on a series of related compounds, to test for biological activity, for their effectiveness as a reagent, for the most appropriate protecting group or their suitability as a synthetic intermediate. A good strategy for such a task would be to use the global minimum from one conformation search as the starting structure for the next conformation search, after the necessary editing. Monte Carlo searches tend to work most quickly if the starting structure is low in energy, and this is a useful method for obtaining a suitable conformation. For many complex molecules, the shape of the global minimum is not easy to guess.

This idea can be taken a step further. All the low energy structures of one conformation search may be closely related to low energy structures of related structures. If all of the minima from one conformation search are mutated to a closely related structure and reminimised, then this may produce a similar result to a completely new conformation search, but much more quickly, because all that is required is the reminimisation of however many structures have already been found, and each of these structures may be close to a minimum before the minimisation algorithm is applied, thus speeding up each step (Goodman and Bueno Saz, 1997; Goodman and Leach, 1997). This strategy appears to work rather well. For example, the structure shown in Figure 4.4 requires a lot of computer time for a complete conformation search. Analogues may give different stereochemistries on epoxidation. How can an analogue be designed to give a particular stereochemistry? A complete conformation search could be performed on each analogue, but this would be very time consuming. A mutation strategy is much more effective, allowing the conformation space of the structures in Figure 4.9 to be explored in the same time as was required for the first two conformation searches. The letters inside the ring list whether each double bond points *down* or *up*, going clockwise around the ring from the top. The figure shows that it would be possible to make a number of analogues of the polyether shown in Figure 4.4 by making these analogues of the structure in the top left corner.

4.6 MOLECULAR DYNAMICS AND SIMULATED ANNEALING

Molecular dynamics, a technique which simulates the movement of a molecule, may be used for conformation searching, particularly follow-

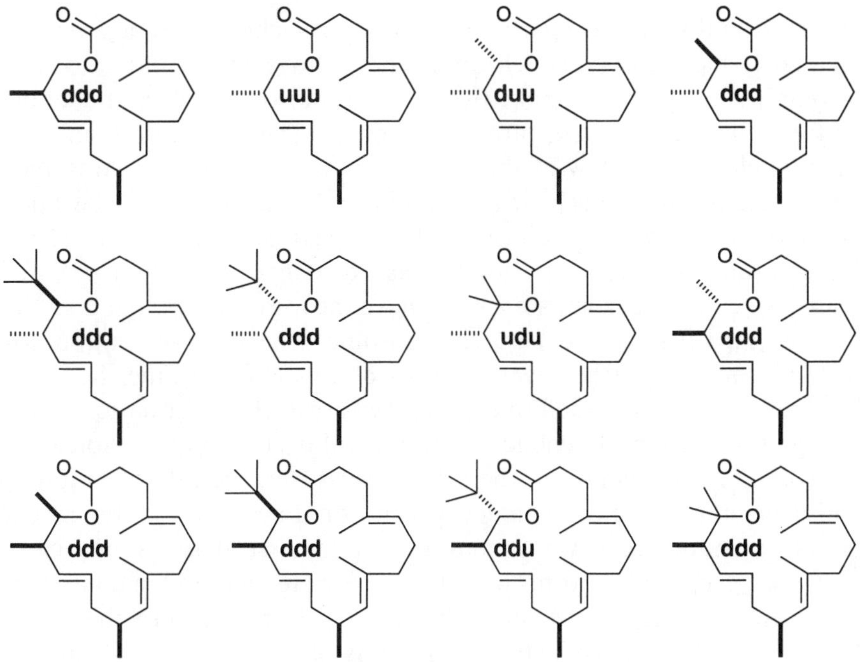

Figure 4.9 *Which face of each double bond is more exposed? Up (u) or down (d)*

ing a protocol called simulated annealing. This is rarely the method of choice for conformation searching of small molecules. It is very useful for exploring the conformation space of larger systems, and this is discussed in Chapter 5. However, it cannot exhaustively search the conformation space of such systems, unlike Systematic Searching or Monte Carlo methods, and so it is not discussed here.

4.7 GENETIC ALGORITHMS

Genetic Algorithms are mathematical techniques for tackling problems in optimisation, drawing ideas from the field of genetics (Holland, 1975). The system which needs to be optimised has to be described in terms of a 'chromosome' which is a string of numbers. For conformation searching problems, this could be a list of torsion angles. For a biological system, this would correspond to a base pair. An initial population of conformations is generated, perhaps by using random numbers or by careful design. This will be the first generation. The next generation is created from these, by mixing and mutating the information in the chromosomes. Too much alteration will destroy the useful information in the previous generation, but too little will make it impossible to explore new areas of conformation space. The mixing is achieved by a *crossover* method, which takes two chromosomes, chops them into two, and recombines the pieces to make two new chromosomes, which must both be the same length as the parents, of course. The mutation is simply a random change in a few of the numbers in a few of the chromosomes. Once these two methods have produced new chromosomes, the best new group of chromosomes is chosen to be the next generation. The criteria for choosing the best may not be easy to define. In a conformation search, one of the chromosomes will have the lowest energy, and so this may be regarded as the best chromosome. The best generation will certainly not simply be several copies of this chromosome, however, because diversity is needed to explore new areas of conformation space. The performance of a genetic algorithm will depend on the choices for the size of each generation, the rate of crossover, the rate of mutation and the method of selecting a new generation. Each of these choices can be made in different ways, and the optimal strategy is not clear. However, genetic algorithms have been shown to be useful in conformation searching (McGarrah and Judson, 1993; Hermann and Suhai, 1995; Meza *et al.*, 1996), particularly in the study of structures for which each entry in the chromosome has a similar significance, such as unbranched alkanes (Nair and Goodman, 1998).

4.8 DISTANCE GEOMETRY

Sometimes a great deal of information is available about a molecule's conformation. This is often the case for proteins which have been studied by NMR, and lists of nuclear Overhauser effects and other data give information about the distance between many atoms. The difficulty is then to invent a conformation which satisfies all, or at least most, of the constraints. A Monte Carlo search could be used, of course, but proteins are so large that this may take many years to converge to a global minimum, and this would not use all of the data which was available. A better procedure which is often used in this sort of situation is called *Distance Geometry* (Crippen, 1981).

It may be very difficult, or impossible, to construct a three-dimensional structure which satisfies all of the experimentally determined constraints. (The experimental data will indirectly give rise to additional constraints, since if atom A is less than 2 Å from atom B, and atom B is less than 2 Å from atom C, atom A cannot be more than 4 Å from atom C.) However, it is generally much easier to construct a model in more than three dimensions, and if the number of dimensions is set to the number of constraints it is straightforward, provided the constraints are consistent. This is rather hard to imagine, but fairly straightforward to write in a computer program. The large number of dimensions gives rise to new degrees of freedom, which are usually filled with random numbers. There are procedures for packing high dimensional structures into fewer dimensions (imagine packing a hinged chain into a flat box), and these can be used to generate three-dimensional structures consistent with the constraints. Because random numbers were used to generate parts of the high-dimensional structures, it may not be possible to find a method to squash the structure into only three dimensions. Many starting structures may have to be tried to obtain reasonable results, and the structures generated may benefit from three-dimensional minimisation.

Chiral centres only exist in three dimensions, and so distance geometry programs must be careful to maintain these features of proteins and other molecules.

4.9 EXPERT SYSTEMS AND ARTIFICIAL INTELLIGENCE

If a chemist sees a two-dimensional representation of a molecule, then a reasonable guess of the three-dimensional structure can be obtained by applying chemical knowledge. For example, benzene rings are flat, cyclohexane rings are often chair shaped, four-membered rings are usually puckered, and so on. This sort of information can be put into a computer program, an expert system, so that three-dimensional structures can be rapidly generated from flat representations. A widely used program which does this is called CONCORD (Pearlman *et al.*, 1987; Rusinko *et al.*, 1989) and other programs have been written (for example, Leach *et al.*, 1990). This method of structure generation is extremely fast, so large databases of three-dimensional structures can be created from two-dimensional data.

The disadvantage of this sort of approach is that only a single reasonable structure is generated. This may not be the global minimum, nor even close in energy to the global minimum. This is not a problem for very rigid systems, but the majority of molecules have conformational flexibility, and a single reasonable conformation may not give a good description of the preferred geometry of the molecule.

4.10 ANALYSIS OF RESULTS

These searches produce a huge amount of information, perhaps many thousands of structures. How can it be interpreted?

The simplest approach is to calculate Boltzmann factors for all of the conformations, and this will give an approximate idea of the relative populations of the different species. The number of conformations which are important will be greater at higher energies (see the graph in Appendix A.3).

4.11 SUMMARY

The prejudiced opinions of a conformation searcher: Everything in Table 4.2 is debatable. Different molecular modellers will have widely differing opinions about the relative values of different techniques. This table is included to let you know one set of prejudices, and mentions the techniques simply in the context of conformation searching. Molecular Dynamics, for example, has other uses, which are described in Chapter 7.

4.12 KEY POINTS

- A molecular mechanics study depends on two questions:
  (i) Is the force field reliable in this situation?
  (ii) Has a complete conformation search been carried out?

Table 4.2 *Methods of conformation searching*

Method	Function	Comment
Minimisation	Downhill to nearest minimum	Finds the nearest minimum
Systematic Searching	In principle, this can find all minima	In practice, this is too slow to be very useful for all but the smallest molecules
Monte Carlo Searching	As Systematic Searching, but faster	This is the method of choice for most small molecules
Rule-based Systems	Generates reasonable structures quickly	No guarantee that the results are representative of the molecules shapes
Distance Geometry	Generates a structure consistent with many constraints	May be the best method, if there are a lot of structural data available
Genetic Algorithms	These algorithms can be used in many ways	Extremely effective for conformation searching, in some situations
Molecular Dynamics	Generates many structures	A good way of finding new structures that are similar to an initial structure
Simulated Annealing	Finds global minimum	If the global minimum is the only conformation of interest, this may be the best method

No study has any validity unless *both* of the questions have been addressed.

- Is a conformation search complete? If the answer is either 'No' or evasive, then any conclusion based on the calculations must be treated with great suspicion.
- The difficulty of a conformation search is very approximately given by the expression Difficulty $\approx 6^{N_t - 5N_r}$ where N_t is the number of rotatable torsion angles, and N_r is the number of rings.
- A conformation search involving more than one molecule will be extremely difficult.

QUESTIONS

Q4.1 Alkanes—again

How difficult is it to do a conformation search on unbranched alkanes? Can you find all the conformations of hexane? What is the longest alkane for which you can find the global minimum? Is the assertion in D3.1 that the global minimum of $C_{18}H_{38}$ is hairpin shaped correct? How important does the extended conformation become as the chain length increases?

Q4.2 Methylcyclohexane

How many possible conformations are there for methylcyclohexane? Are the minima restricted to three conformations considered shown in Figure 4.2 (chair with axial and equatorial methyl, twist boat) or are there more?

Q4.3 Very sterically hindered cyclohexanes

What is the preferred conformation of hexamethylcyclohexane and of hexa-*iso*-propylcyclohexane (Figure 4.10)?

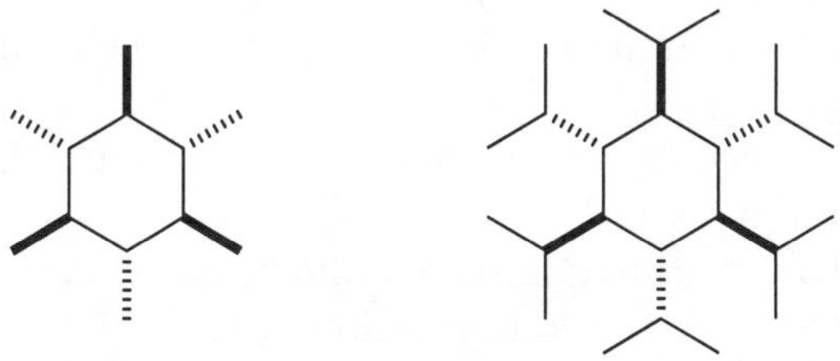

Figure 4.10 *Hexa-substituted cyclohexanes*

Q4.4 How large a ring is needed to fit a *trans* double bond?

Cyclopentene and cyclohexene have *cis* double bonds. A *trans* bond would not really fit into a five-membered ring, but there should not be a problem if the ring is made large enough. What is the smallest ring which can fit a *trans* bond (Figure 4.11)?

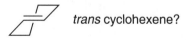 *trans* cyclohexene?

Figure 4.11

Q4.5 Molecules with one low energy conformation

What are the preferred conformations of the molecules shown in Figure 4.12? Can you see why only one conformation is favoured in each case?

Figure 4.12 *Molecules with only one low energy conformation*

Q4.6 Various molecules

Figure 4.13 shows six molecules.

(a) For which of these molecules will solvent effects be most important?

(b) For which are reliable force field parameters unlikely to be available?

(c) For each molecule: which torsion angles need to be rotated in a conformation search?

(d) Why will a conformation search of *6* be particularly difficult?

(e) What is the preferred conformation of *1*?

(f) Is the backbone of *3* likely to be linear? What factors control its preferred conformation?

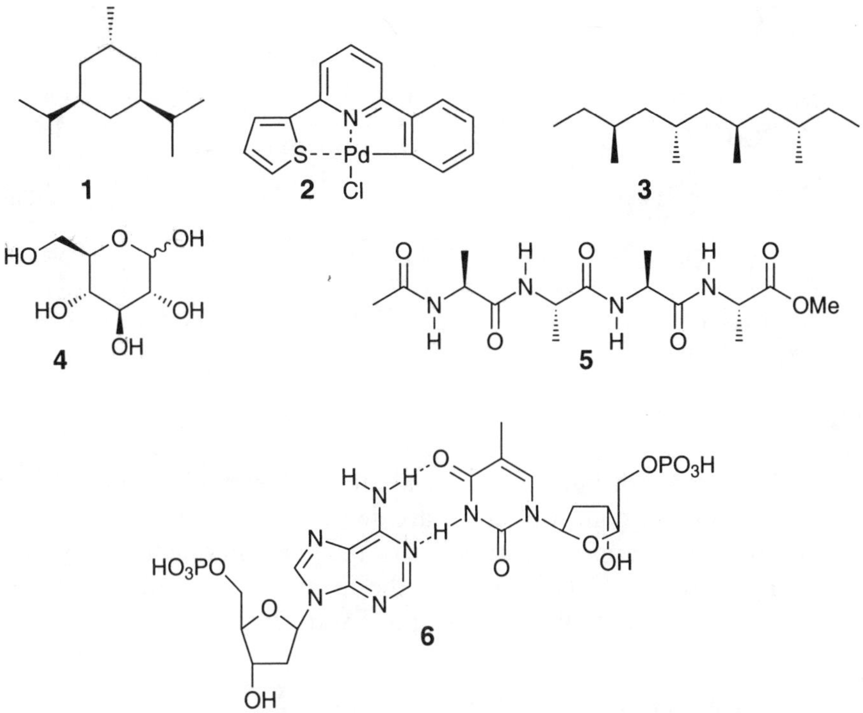

Figure 4.13

Q4.7 Erythromycin

Erythromycin A is an important antibiotic, with the complicated structure shown in Figure 4.14.

Intermediates in the synthesis of erythromycin are shown in Figure 4.15.

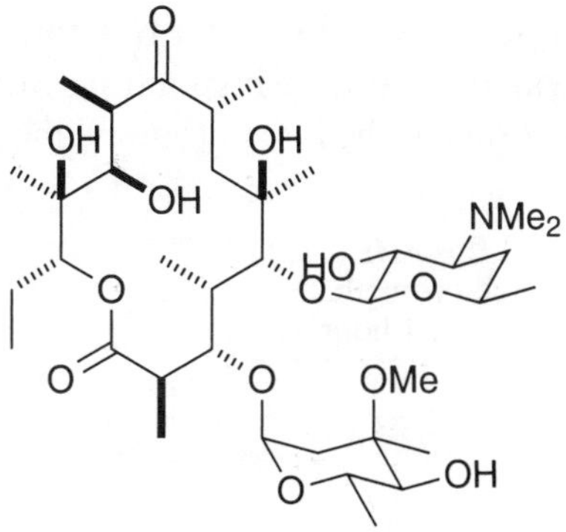

Figure 4.14 *Erythromycin A*

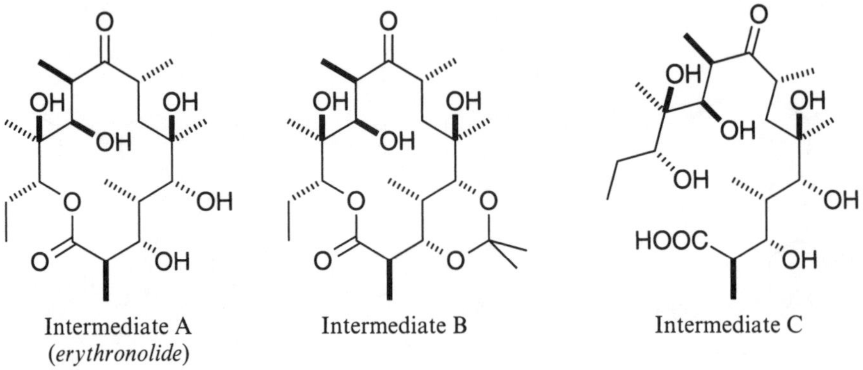

Intermediate A   Intermediate B   Intermediate C
(erythronolide)

Figure 4.15

(a) Arrange erythromycin itself and all the intermediates in order of difficulty for conformation searching.

(b) What is the approximate shape of the final intermediate, erythromycin seco-acid (which lacks the macrolide ring all the others contain) likely to be *in vacuo* and in water?

Q4.8 Account for the selectivity of a rearrangement

The diol shown in Figure 4.16 undergoes a pinacol rearrangement when treated with a Lewis acid ($Et_2O \cdot BF_3$). The pinacol rearrangement could form a mixture of two different ketones, but experiment shows that only one is formed. How can molecular mechanics be used to analyse this reaction, and to rationalise the observed selectivity?

Here is a possible course of action:

(a) *Assume* that the reaction is thermodynamically controlled. In this case, the ratio of the products will be controlled by the relative energies of the products.

(b) Find the energies of the low energy conformations of the two ketones, and calculate the corresponding Boltzmann factors. The relative proportions of the two ketones will be given by the ratio of

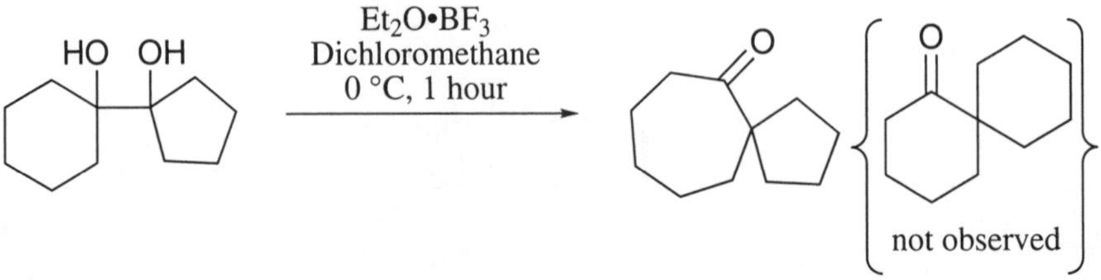

$$Et_2O \cdot BF_3$$
Dichloromethane
0 °C, 1 hour

not observed

Figure 4.16 *Pinacol rearrangement*

their Boltzmann factors. Is the experiment consistent with the premise that the reaction is thermodynamically controlled?

(c) *Assume* that the reaction is kinetically controlled. The size of the energy barrier that must be overcome to form the products must be estimated. In order to do this, note that the carbonium ion formed by removal of one of the hydroxyl groups will be similar in structure to the transition state of the reaction. Calculate the relative energies of the two carbonium ions. What ratio of products is expected now?

(d) Are the results consistent with the experiment?

(e) Do the results suggest that the product which is not formed under these condition could be made under different reaction conditions?

Q4.9 Cram's Rule

Non-chelating nucleophilic attack on the two diastereomeric aldehydes (*syn, anti*) shown in Figure 4.17 gives opposite selectivity, even though Cram's Rule suggests that the selectivity should just depend on the α-centre. Can molecular mechanics be used to suggest an explanation for this surprising selectivity?

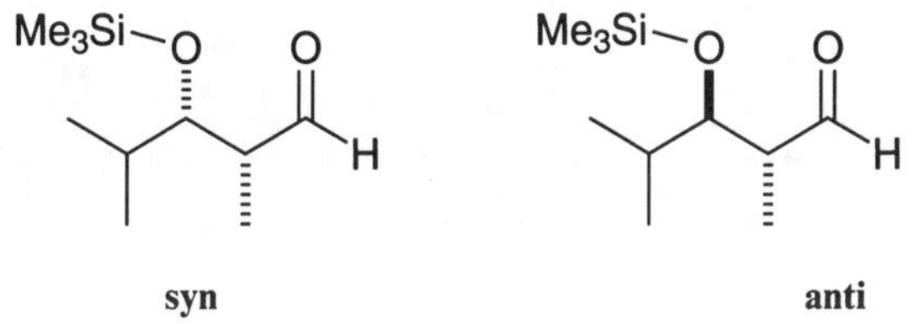

syn   anti

Figure 4.17

Q4.10 Explain the selectivity of a reaction

Alkylation of the ketone shown in Figure 4.18 gave essentially one product in high yield (Begley *et al.*, 1988). LHMDS is lithium hexa-

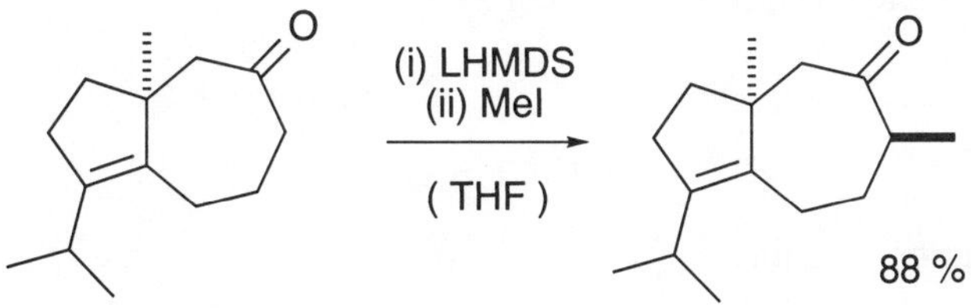

(i) LHMDS
(ii) MeI

(THF)

88 %

Figure 4.18

methyldisilazide, a strong, non-nucleophilic base. Explain the selectivity. Would you have been confident enough to predict the result, if the answer had not been given?

Q4.11 Selectivity of elimination

Why does the elimination shown in Figure 4.19 occur selectively to form the non-conjugated product? Might other double bond isomers be formed?

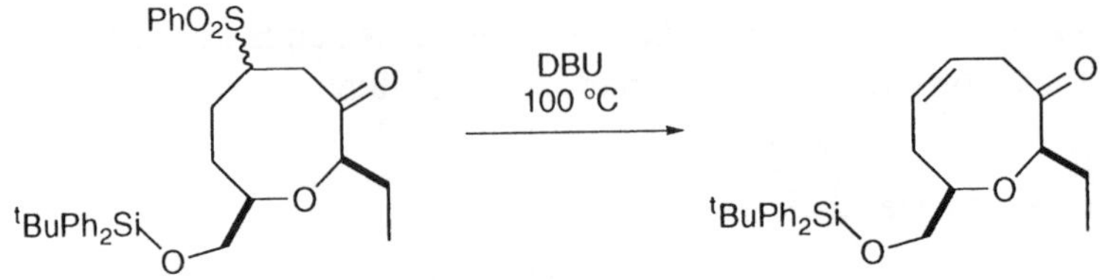

Figure 4.19

Q4.12 Selectivity of hydrogenation

Is the hydrogenation shown in Figure 4.20 likely to give the *S* product, the *R* product, or a mixture?

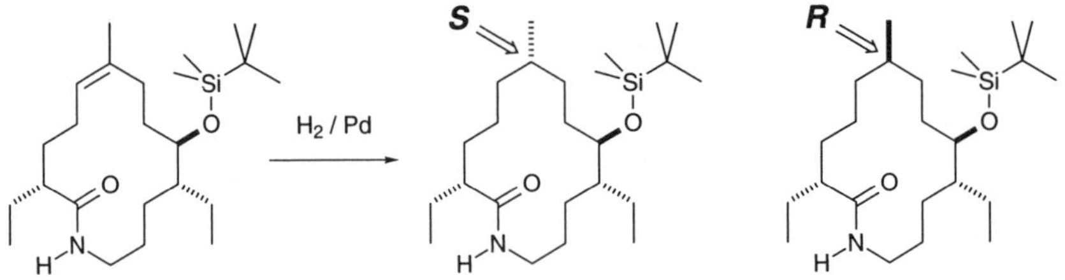

Figure 4.20

Q4.13 DNA, RNA and enzymes

Is it possible to do a conformation search on a DNA double helix or on an RNA strand? What about an enzyme?

DISCUSSION

D4.1 Alkanes—again

The rule of thumb that difficulty is proportional to six to the power of the number of rotatable torsion angles suggests that the difficulty of a

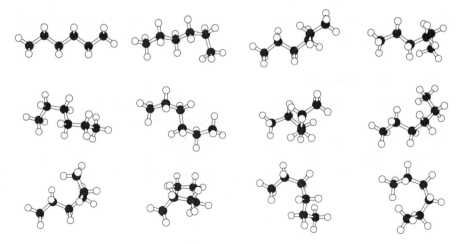

Figure 4.21 *All the minima for hexane*

conformation search on unbranched C_nH_{2n+2} should be about 6^{n-3}. Hexane is straightforward, and has rather less than $6^3 = 216$ distinct conformations. Once all the conformations which are equivalent by symmetry are removed, only twelve remain (Figure 4.21, and Plate 2).

The global minimum energy conformation for hexane is the extended chain (top left of the figure). In discussion D3.1 it was stated that the global minimum of $C_{18}H_{38}$ is hairpin shaped, for MM2*. This molecule is probably too large for a comprehensive conformation search in a reasonable length of time, and the basis for this assertion is a consideration of the energetic cost of the twist in the chain, compared with the favourable interaction of each methylene unit. It is conceivable that another conformation will be even lower in energy. Monte Carlo searches work reasonably well for this sort of system, but genetic algorithms appear to be more successful in this particular case.

The importance of the extended conformation may be estimated from its Boltzmann factors. The energetic cost of making one torsion angle *gauche* rather than *trans* is about $4\,kJ\,mol^{-1}$, corresponding to a Boltzmann factor of about one fifth at room temperature. However, there are many ways of twisting a single torsion angle. Three are illustrated in Figure 4.21, but this neglects conformations related by symmetry, so there are in fact ten conformations $4\,kJ\,mol^{-1}$ above the extended chain conformation, each with a population of about one fifth of the extended chain. The extended conformation cannot, therefore, represent more than a third of the conformations present. In fact, it will be less than this, because doubly twisted conformations will also be important. Medicinal chemists often make analogues by substituting a methyl group with an ethyl group, then a propyl, and then stop (methyl,

ethyl, propyl, futile). However, a hexane chain will not obediently stay extended and out of the way, so this rule of thumb may be misleading.

D4.2 Methylcyclohexane

According to MM2*, there are five minima in the methylcyclohexane potential energy surface, disregarding those related by symmetry. The lowest two are both chairs, and the other three are all twist boats with different methyl positions (Figure 4.22). Equatorial methylcyclohexane is 7.4 kJ mol^{-1} lower in energy than axial methylcyclohexane. The three boat forms have rather similar energies.

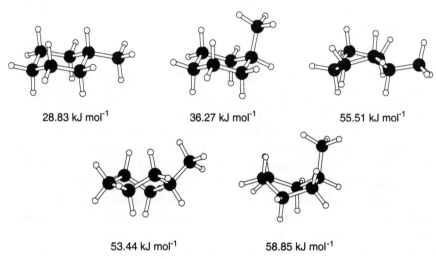

28.83 kJ mol^{-1} 36.27 kJ mol^{-1} 55.51 kJ mol^{-1}

53.44 kJ mol^{-1} 58.85 kJ mol^{-1}

Figure 4.22 *The five minima of methylcyclohexane, ordered by energy*

D4.3 Very sterically hindered cyclohexanes

A first guess might well be that both compounds (Figure 4.23) would prefer the conformation with all the substituents equatorial. This is the

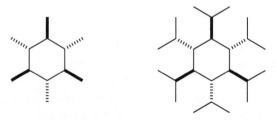

Figure 4.23 *Hexa-substituted cyclohexanes*

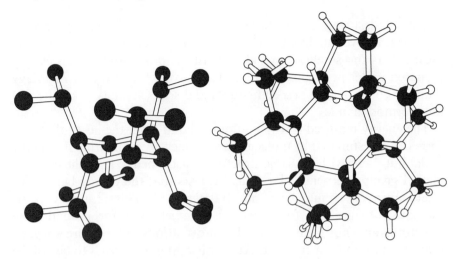

Figure 4.24 *Conformation of hexamethyl-*iso-*propylcyclohexane*

case for hexamethylcyclohexane. The all-equatorial conformation is $30\,kJ\,mol^{-1}$ lower in energy than the all axial conformation. This is much less than six times the energy difference between equatorial and axial methylcyclohexane (previous question). There are also many boat conformations, of course, and these lie at intermediate energies.

Hexamethyl-*iso*-propylcyclohexane's global minimum conformation (Figure 4.24) has an all axial arrangement of sidechains! The all equatorial conformation is much higher in energy, above many boat structures, because of the unfavourable interaction of the sidechains in the equatorial conformation.

D4.4 How large a ring is needed to fit a *trans* double bond?

The global minima from conformation searches on *cis* and *trans*-cycloalkenes with various ring sizes are listed in Table 4.3.

Table 4.3 *Global minimum energies of* cis *and* trans *cycloalkenes* $(kJ\,mol^{-1})$

Ring size	Energy of cis form	Energy of trans form
6	17.869	173.209
7	41.90	157.858
8	56.59	97.49
9	74.84	90.69
10	69.89	75.46
11	72.38	67.85
12	70.80	66.38

The smallest ring for which the *trans* form is lower in energy than the *cis* form is cycloundecene. However, *trans*-cyclooctene is generally regarded as the smallest cycloalkene with a *trans* bond which can be prepared under normal conditions. This is very much higher in energy than the corresponding *cis* compound, and so its stability is kinetic and not thermodynamic.

The energy required to twist a double bond from the *cis* to the *trans* form can be estimated by molecular mechanics, by comparing the energy of flat but-2-ene with the same structure constrained to have the terminal methyl groups perpendicular. This suggests that the energy barrier for interconversion should be about $100\,\text{kJ}\,\text{mol}^{-1}$. This is a very crude estimate, because the force field is parameterised for ground state structures and not such distorted conformations. This figure suggests that the amount of energy released by changing from *trans* to *cis* will be sufficient to overcome this barrier for six- and seven-membered rings, but not for eight-membered rings, which is consistent with experimental observations. However, the uncertainty in the measurement of the barrier for interconversion is so large that the possibility of a *trans* bond in a seven-membered ring should not be ruled out on the basis of these calculations.

Within the table, the energy of the rings containing *trans* double bonds decreases with increasing ring size, whilst the energies of the rings containing *cis* double bonds increases and then falls again. This is presumably because the favourable van der Waals interactions across the rings make larger rings lower in energy. The trend for the larger rings suggests that *trans* bonds will be favoured for very large rings, as they are for chains.

D4.5 Molecules with one low energy conformation

The global minimum conformations of the two compounds shown in Figures 4.25 and 4.26 are much lower in energy than the lowest local minima, because any change to the structure leads to a large increase in energy.

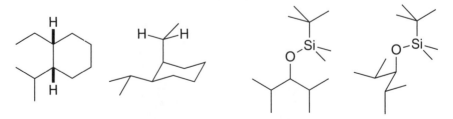

Figure 4.25 *Global minimum energy conformations*

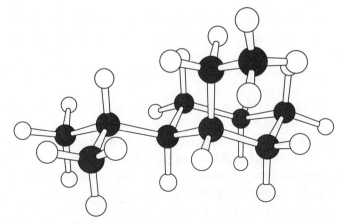

Figure 4.26 *Another view of the global minimum conformation of 2-ethyl-iso-propylcyclohexane*

D4.6 Various molecules

(a) *For which of the molecules in Figure 4.27 will solvent effects be most important?*

Those with important non-bonded interactions—so **4, 5, 6** (hydrogen bonds). The anomeric centre's preferred configuration in **4** will

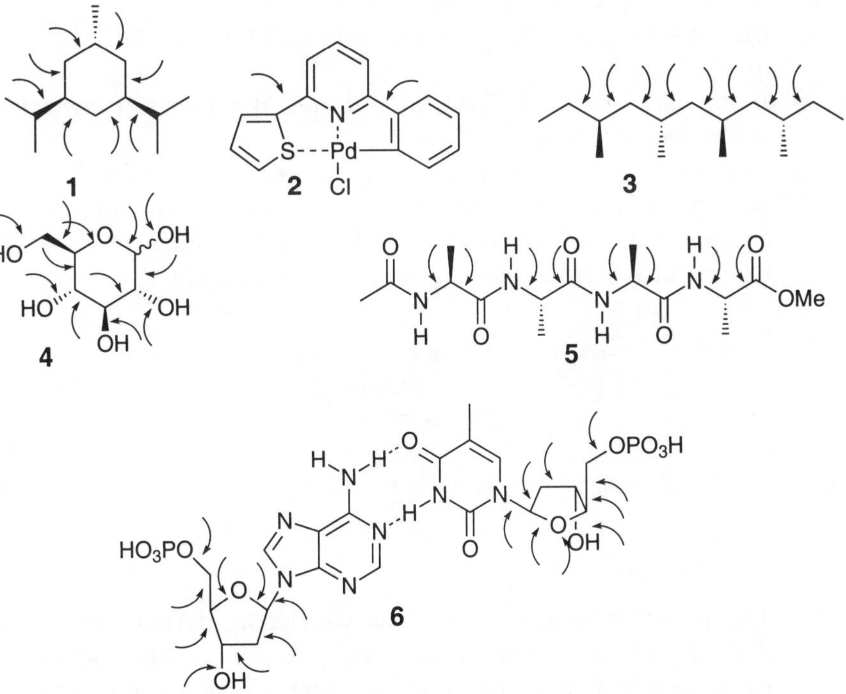

Figure 4.27 *Arrows mark bonds which must be rotated in a conformation search*

also be affected. Solvent effects will play a role in the conformation of all molecules, but this will be a very small factor in molecules like **1**.

(b) *For which are reliable force field parameters unlikely to be available?*

Force field parameters for palladium are not readily available, so it will be hard to do a conformation search on **2**. If parameters are available for this system, then it is important, as ever, to consider how reliable the parameters are likely to be. Parameters for phosphorus are only available for some functional groups. Structure **6** is a DNA fragment and DNA has been studied in great detail, and so parameters will be available.

(c) *For each molecule: which torsion angles need to be rotated in a conformation search?*

Marked on Figure 4.27. Hydroxyls need to be rotated. Methyl groups do not. It is probably reasonable to assume that the amide bonds in **5** stay Z in all the low energy conformations.

(d) *Why will a conformation search of 6 be particularly difficult?*

Structure **6** is two molecules, which must be moved relative to each other. The global minimum structure may have the π-systems stacked, rather than the hydrogen bonded conformation shown.

(e) *What is the preferred conformation of 1?*

Figure 4.28 is the only conformation which avoids all 1,5 diaxial interactions.

(f) *Is the backbone of 3 likely to be linear? What factors control its preferred conformation?*

The preferred conformation of **3** is twisted (Figure 4.29). This has no significant effect on the bond strain, angle strain and torsional energies, but increases the favourable non-bonded interactions, by placing atoms close, but not too close, to each other.

Figure 4.28 *Global minimum conformation of 1*

D4.7 Erythromycin

(a) The first three molecules are cyclic, which considerably reduces the difficulty of the conformation search. A conformation search on Intermediate B is easiest because there are two rings. Erythromycin A (Figure 4.14) is harder than Intermediate A (Figure 4.15)

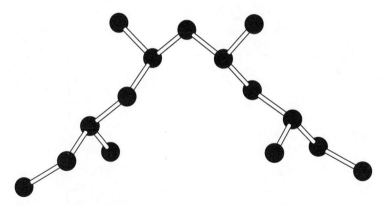

Figure 4.29 *Global minimum conformation for 3. This conformation has more favourable van der Waals interactions than the extended form*

because it is larger, even though it has more rings. The order of difficulty can be determined by applying the rough guide: Difficulty $\propto 6^{(N_t - 5N_r)}$ where N_t is the number of rotatable torsion angles, and N_r is the number of rings. The sugars in Erythromycin A and the six-membered ring in Intermediate B probably have a strong preference for one conformation, so it may be reasonable to avoid rotating the torsion angles in these rings.

(b) In vacuum, the extended form of the seco-acid is preferred, with a line of hydrogen bonds to hold the molecule approximately linear. If a solvent model is used, a folded form is preferred with the terminal acid and alcohol close to each other, as the strength of the intramolecular hydrogen bonds is weakened and there is an energetic advantage in minimising the molecule's surface area.

Plate 3 shows the 20 lowest energy conformations of Intermediate A (erythronolide).

D4.8 Account for the selectivity of a rearrangement

This diol undergoes a pinacol rearrangement (Figure 4.30) when treated with a Lewis acid ($Et_2O\cdot BF_3$). The pinacol rearrangement could form a mixture of two different ketones, but experiment shows that only one is formed (Sands, 1994).

Calculation of the relative energies of the two products gives an energy of $82.0\,\text{kJ}\,\text{mol}^{-1}$ for the global minimum of the 6,6 product, which is not observed, and $116.7\,\text{kJ}\,\text{mol}^{-1}$ for the 7,5 product which is formed. This strongly suggests that the reaction is not thermodynamically controlled.

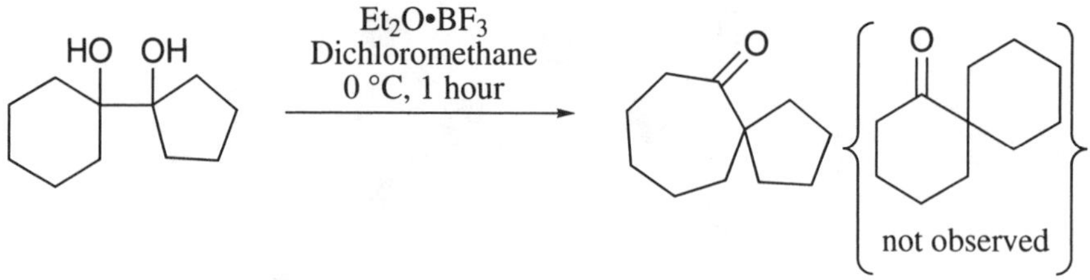

Figure 4.30 *A pinacol rearrangement*

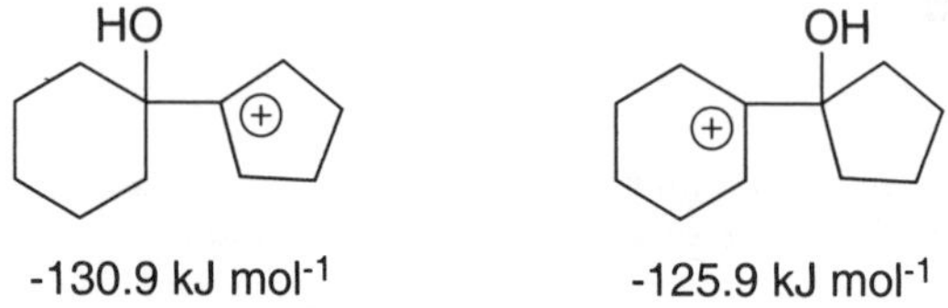

-130.9 kJ mol⁻¹   -125.9 kJ mol⁻¹

Figure 4.31 *The energies of the global minima of the cations*

Calculation of the relative energies of the two cations (Figure 4.31) suggests that the cation in the five-membered ring is lower in energy, and this is similar to the transition state leading to the observed product. Therefore, the experimental observations are consistent with the calculation and the assumption that the reaction is kinetically controlled.

This calculation could be criticised on two grounds. (i) Only the energies of the global minima have been used, rather than the energies of all low energy conformations. In this case, this makes little difference, but this was only clear when all the low energy conformations had been calculated. (ii) The parameters for the cations are unlikely to be very reliable. However, they may be reliable enough for this comparison. The only way to be sure would be to find more experimental data.

D4.9 Cram's Rule

Non-chelating nucleophilic attack on the two diastereomeric aldehydes (*syn, anti*) shown in Figure 4.32 gives opposite selectivity, even though Cram's rule states that the selectivity should just depend on the α-centre. MacroModel is used to suggest an explanation for this, based on the MM2* force field.

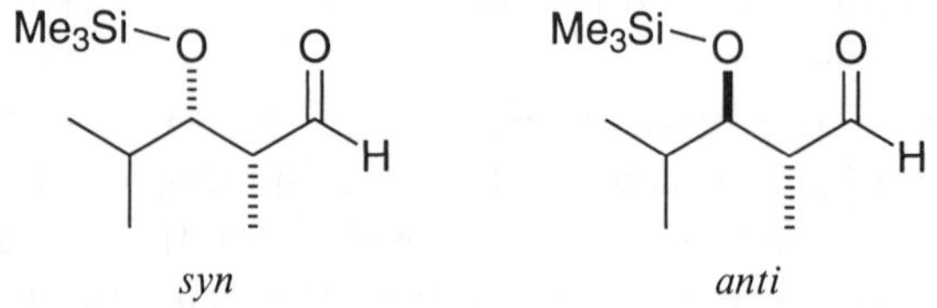

syn   anti

Figure 4.32

Table 4.4 *Lowest energy structures of diastereomeric aldehydes shown in Figure 4.32*

syn				anti			
Energy (kJ mol^{-1})	Boltzmann factor[a]	Times found	Accessible face	Energy (kJ mol^{-1})	Boltzman factor[a]	Times found	Accessible face
59.34	1.00	8	re	62.90	1.00	1	si
59.34	1.00	9	re	62.90	1.00	5	si
59.34	1.00	11	re	62.90	1.00	6	si
61.75	0.23	6	re	62.93	0.98	11	si
61.75	0.23	3	re	62.93	0.98	9	si
61.75	0.23	2	re	62.93	0.98	3	si
63.56	0.08	9	si	63.28	0.80	11	re
63.56	0.08	9	si	63.28	0.80	14	re
63.56	0.08	10	si	63.28	0.80	7	re

[a] Boltzmann factors calculated at $-78\,°C$, and all scaled so that the largest Boltzmann factor is 1.00 for each molecule.

A Monte Carlo conformation search must be performed on both structures. The lowest energy structures are listed in Table 4.4.

The conformation searches found the low energy structures in groups of three, each with identical energies. The reason for this can be found by comparing the conformations. The conformations within each group are identical, except for the trimethylsilyl group, which has been rotated through 120°. The computer did not realise that these conformations were equivalent. It is reasonable, therefore, to consider each of these groups of three as being a single conformation.

All of the lowest energy structures were found several times each (after uniting the groups of three) and so there can be some confidence that the conformation search was complete in both cases. In the case of the *syn* conformation, the Boltzmann factor of the third lowest energy structure is only 0.08, and so considering only the three lowest energy structures is reasonable. For the *anti* form, the three lowest energy structures all have similar energies and Boltzmann factors, and so the table does not contain enough data to analyse the conformations fully. Higher energy conformations should also be included.

The conformations of the global minima are shown in Figure 4.33. It is clear that the *re* face of the *syn* isomer is much more accessible than the *si* face (see Appendix A.6). For the *anti* isomer, the situation seems less clear cut, but the *si* face appears to be the more accessible. This is consistent with the experimental data (Paterson *et al.*, 1994). The *syn* isomer shows high selectivity for attack on the *re* face, whilst the *anti* isomer prefers attack on the *si* face, but with much lower selectivity.

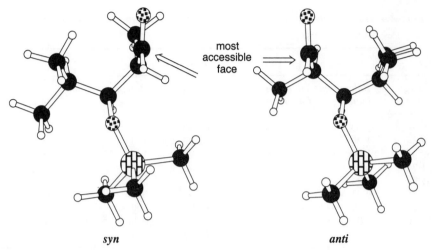

Figure 4.33 *The most accessible face for attack of the global minima*

D4.10 Explain the selectivity of a reaction

Two questions must be addressed to account for the experimental results (Begley *et al.*, 1988): (a) the regioselectivity of enolisation and (b) the stereoselectivity of addition to the enolate.

Good parameters were not available for the enolates, so alkenes were used as models for them, since these should have similar geometries and good parameters are available. Conformation searches were performed on all the molecules in Table 4.5 and the energies of the global minima are listed. The conformation searches showed that there were only a few low energy structures in each case, so it is reasonable to use the global minima only to assess the selectivity of the reaction.

(a) The calculation shows that the thermodynamic product of the enolisation reaction is on the right hand side of the ketone (ratio of alkenes 5.6 : 1 at RT; 13 : 1 at −78 °C). The reaction may well be kinetically controlled, so the hindered base approaches the less hindered side of the ketone. In this case, the kinetic and thermodynamic products are the same.

(b) The preferred conformation of the enolate presents the upper face of the double bond for electrophilic attack, so the observed diastereoselectivity follows. The calculations show that this product would also be the thermodynamic product of the reaction.

The analysis of the reaction could be improved in a number of ways: (i) Use all the conformations of the reactants, rather than just the global minima to assess the selectivity. (ii) Use an enolate model instead of an

Table 4.5 *Global minimum energies* (kJ mol^{-1}) *for various structures*

Structure	Energy	Comment
	83.1	
	90.8	the upper face of the double bond is most open to attack by an electrophile
	86.5	
	88.5	
	94.1	
	94.9	
	100.2	

alkene. The lithium ion would have to be included in the model, and some parameters are available for this, although new ones may need to be developed.

D4.11 Selectivity of elimination

Simplified forms of the molecule were chosen for the conformation searches (Figure 4.34). A list of low energy conformations is given in Table 4.6. The reaction conditions suggest that the reaction will go under thermodynamic control, and so calculation of the relative energies of the products will be enough to predict the selectivity of the reaction.

The results of the conformation search show that the non-conjugated

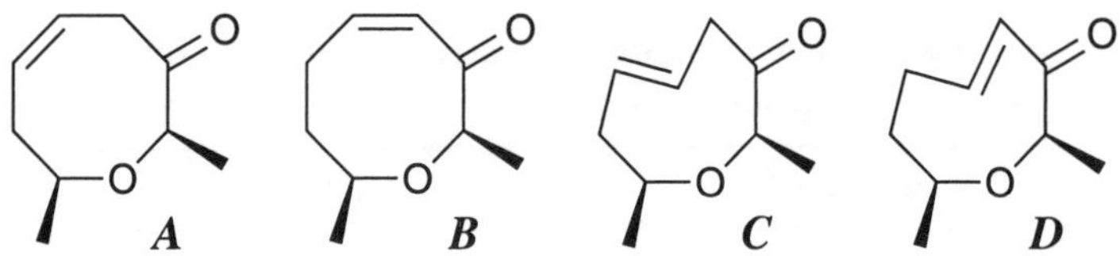

Figure 4.34 *The structures chosen for the conformation search*

Table 4.6 *Minimum energies* $(kJ \, mol^{-1})$ *for isomers shown in Figure 4.34*

Isomer	Conformation			
	1	2	3	4
A	−53.33	−48.11	−44.67	−43.40
B	−20.10	−19.01	−16.70	−10.11
C	5.08	11.01	15.84	
D	45.57	51.96		

form of the enone is strongly preferred, in line with the experimental result (Mujica *et al.*, 1996), but against normal expectation. The *Z* double bonds are strongly preferred over the *E* double bonds.

D4.12 Selectivity of hydrogenation

Molecular models suggested that the hydrogenation should go as shown in Figure 4.35, and it did, with >95% selectivity (Houri *et al.*, 1995). A conformation search of the system gives much greater confidence that only one face of the double bond is accessible, and this is the one which gives the desired *S* stereochemistry.

D4.13 DNA, RNA and enzymes

Conformation searches on DNA, RNA and enzymes are likely to require too much computer time to be worth considering.

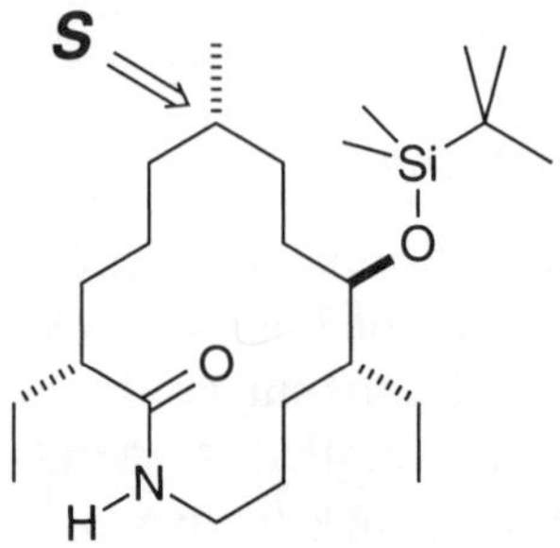

Figure 4.35

CHAPTER 5

Molecular Dynamics

'Eppur si muove'
(attributed to Galileo Galilei)

5.1 ANALYSIS OF FORCE FIELD MODELS

Previous chapters have introduced the use of force fields to describe molecular structure. A force field is a description of a multidimensional energy surface, which is a complete description of the molecule in all its conformations. The reliability of this description can, and should, be questioned, but in this chapter it will be assumed that force fields are reasonably trustworthy. The problem that must be faced is how to extract useful information from a force field. The quantity of data that is available from a force field is so great that this is a very difficult problem.

Chapter 4 discussed one way to get useful information from force fields: making a list of minima in the potential surface of a molecule. This will give some idea of the conformations that the molecule will take up, but it assumes that molecules can be treated as collections of static conformations. This is not the case. Molecules are always moving, even at the absolute zero of temperature. Molecular dynamics simulates this movement.

A conformation search is one way of extracting useful information from a force field, but there are cases where this does not give all the information which might be expected. For example, the molecule in Figure 5.1 (Guella *et al.*, 1997) is a particular challenge for conformation searching, because it can exist in two slowly interconverting conformers. Could this have been predicted? A conformation search would probably find both the low energy conformers, but would not give the information that they were slowly interconverting. This information could be obtained by a long molecular dynamics simulation.

To take an example from biological chemistry, consider a strand of DNA containing twelve base pairs (Figure 5.2). It is possible to calculate the difficulty of doing a full conformation search on a system like this,

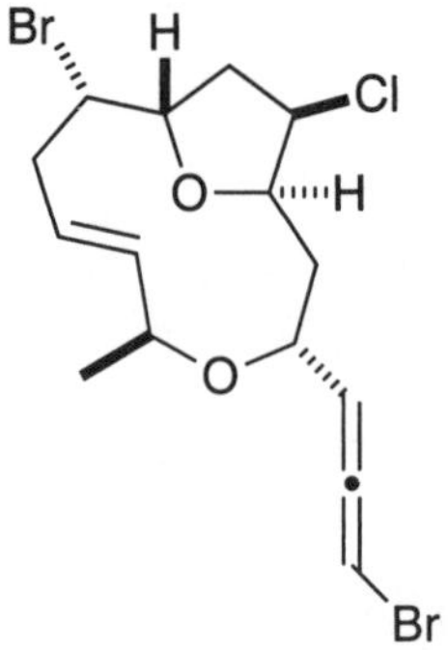

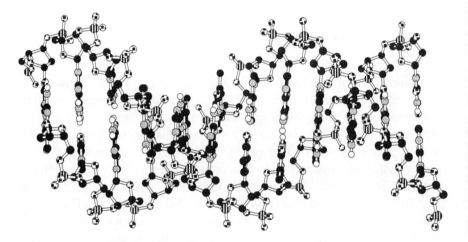

Figure 5.1

Figure 5.2 *B-DNA (CGCGAATTCGCG)$_2$*
(Rao and Kollman, 1990)

using the approximate expression in Equation 4.2. The answer is about ten to the power of eighty, which is thought to be the number of particles in the universe. There is no possibility of doing a full systematic or Monte Carlo conformation search on a system of this size. There is no difficulty in minimising it, however. Molecular dynamics provides a useful method of getting quantitative information which can be related to experiments for systems of this type.

The same sort of problem applies to studying molecules in solution. A very large number of solvent molecules are required to solvate even the smallest of systems (potassium ions require about one hundred water molecules, for example (Figure 5.3), and larger molecules require many more). Exhaustive conformation searching is not a useful approach in

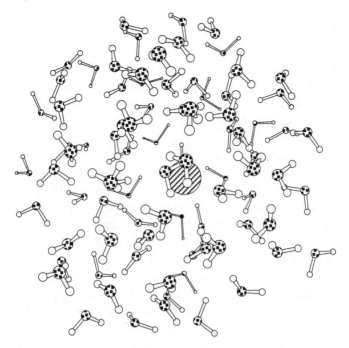

Figure 5.3 *A potassium ion surrounded by about 100 water molecules*

this sort of situation. Would it be useful, however, to find out all the possible arrangements of the solvent? The solvent's effect on the solute may be the key question. If a sufficient number of arrangements of the solvent molecules are considered, so that a representative sample has been found, this may be enough to provide useful data. Molecular dynamics provides a way of producing such representative sets of conformations.

If a potential surface is likened to a mountainous landscape, then the valleys are the most interesting parts of it, because these are where a molecule will spend most of its time. The easy routes between the valleys may also be important, but the mountain peaks are not, because the energy required to reach the summits may not be available. Force fields are parameterised with the valleys in mind, and so they will usually provide a better description of the valleys than of the mountain tops. Because only limited energy is available to molecules, many mountains may be too high to climb. If a mountain is inaccessibly high, there is no interest in discovering whether it is twice as high as the highest point that can be reached, or ten times as high. A description of this landscape which describes only the parts which are accessible, and how to get from one part to another without climbing too high, is all that is needed for a complete description of the interesting properties of the system.

5.1.1 The Shape of the Landscape

Conformation searching, if it is taken to its limit, will provide a list of the global minimum and all the local minima conformations that a molecule may attain. The populations of these conformations may be described by a Boltzmann distribution, but this requires two assumptions: first, the conformations are in equilibrium; second, the energy barriers to inter-conversion are high.

The first assumption is necessary for the populations of the minima to be distributed as a Boltzmann distribution. The second is needed because it is assumed that only the minima are important. If the barriers for interconversion are low, then the molecules may spend most of their time moving between minima, instead of being in minima.

These assumptions are not totally compatible with each other. The four graphs on the left hand side of Figure 5.4 present four different fragments of potential surfaces. It is obvious that the properties of the molecule which correspond to this part of the surface will change gradually as the barrier between the two wells decreases. However, as far as the Boltzmann distribution analysis is concerned, the top three

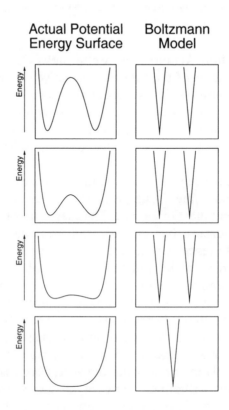

Figure 5.4

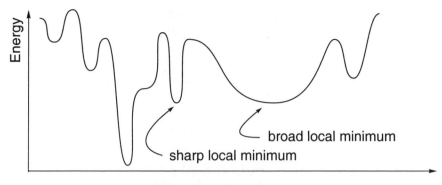

Figure 5.5

graphs are all identical to two steep potential wells, as is illustrated on the right of Figure 5.4 and the lowest graph is identical to one potential well. There is a sharp change in behaviour as the potential barrier alters from being very small to being zero, and this is not a reasonable physical picture. The error does not come from the force field, but from the analysis of the force field.

A potential energy surface may have minima which are broad and minima which are sharp (Figure 5.5). The two local minima which are marked have the same energy (Molecular Mechanics energy, E_{MM}, or internal energy U). However, they will have rather different free energies (G) because the broad minimum allows more freedom to move. A normal conformation search will not be able to detect the difference between these minima.

5.1.2 Constraining Exploration

There is a second problem with the conformation-searching analysis of a force field. Often, only a very small part of a potential energy surface is interesting, for example that part which corresponds to the bound state of a protein and its substrate. More of the potential surface will be the unbound states of the pair, because there are so many more ways that the two molecules can be arranged apart rather than together. It is very hard to limit a conformational search to consider only bound states. If some knowledge of the binding conformation is available, perhaps from nOe data, then this could be used to constrain certain bond lengths, and would place a sharp limit on the search (Figure 5.6). The limits which are often important, however, are defined by potential barriers, and these cannot be exactly reproduced by simple distance constraints. Molecular

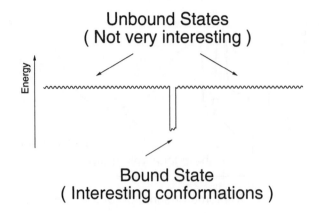

Figure 5.6

dynamics provides a way to do this, at the cost of a large amount of computer time and various new errors.

5.2 HOW CAN MOVEMENT BE MODELLED?

A force field model of a molecule behaves rather like a collection of balls on springs. Such a mechanical model tends to wobble about and this is usually seen as a disadvantage. However, even though it is annoying, it is arguably a better model of a real molecule when it is moving than when it is still, because real molecules wobble about too. It is possible to calculate how a model will move, and it is also possible to calculate how a force field model of a molecule will oscillate. This process is called a molecular dynamics simulation.

The movement of one of the atoms in the molecules is affected by the atoms around it, which in turn are affected by the first atom. This type of mathematical problem, coupled differential equations, has been studied in great detail. In general, it is not possible to solve these equations analytically, but approximate solutions can be found fairly easily by computational methods.

5.2.1 The Molecular Dynamics Approach

A molecular dynamics simulation begins by giving each atom in a molecule some kinetic energy. This makes the molecule move around, and it is possible to calculate how it moves by solving the Newtonian equations of motion. This is done by analysing what the molecule is doing, and using this to predict what will be happening a very short time in the future. The calculation is a difficult one, and it takes a lot of computer power to simulate how a molecule will move for a few picoseconds. This is a much shorter time then is generally of interest, but useful information can be gained from short simulations.

Molecular dynamics mimics the way a molecule actually explores its conformational space, rather than trying to get a picture of the whole of the conformational space, as conformational searching methods do. This is an advantage if the conformational space is so large or so intricate that conformational search methods cannot be continued until a complete picture begins to emerge. Molecular dynamics, therefore, is particularly suited to studying protein conformations, or other large molecules for which there is incomplete structural data.

5.2.2 Parameters for Molecular Dynamics

A molecular dynamics simulation needs

- A starting structure
- The temperature
- The step size
- The length of the run

5.2.2.1 Starting Structure. A molecular dynamics simulation must begin with a sensible geometry for the structure of interest, and this may be difficult to obtain. Typically, an approximate model will be built using any experimental data that are available, maybe nOe or crystal structure data, and this will then be minimised (almost certainly to a local minimum rather than the global minimum) so that all the bond lengths and bond angles have sensible values. One of the methods of conformational searching previously described might be used to find a lower energy structure. It is not necessary to start with the global minimum, or even a structure close to the global minimum, but if the starting structure is too strained, errors in the molecular dynamics simulation may accumulate too rapidly, and the structure may 'blow up'.

5.2.2.2 Temperature/Energy. The amount of energy the structure is given depends on the temperature of the system. The energy is divided between movement and potential energy, just as a ball on a spring will be moving quickly when the spring is relaxed, and stationary when the spring is fully compressed or fully extended.

The energy is divided between the atoms so that each atom gets more or less the same energy. If all the energy were given to one atom, then that atom would probably escape from the molecule, and this is not a situation that force fields are good at coping with.

Errors in the calculation mean that it is necessary to check the energy after every step, and adjust it so that the system does not get carried away! This is due both to the errors in the numerical solutions of Newtonian equations, and to the inability of force fields to predict the properties of very distorted structures. (For example, force fields may allow chiral centre inversion, simply by pulling hard on the bonds of the centre, or chains to pass through each other so that the linked rings of catenanes may become unlinked.)

5.2.2.3 Step Size.   The calculation of how the molecule will move works by calculating what the molecule will be doing a very short time in the future. This short time has to be much shorter than the shortest time in which anything interesting can happen to the molecule. One of the problems with solving coupled differential equations is finding out how short a time is necessary. This is called the characteristic time for the system. The fastest thing that can happen to a molecule is an electronic transition. However, molecular mechanics ignores electrons, so these need not be considered. The fastest mechanical change to a molecule is the property which determines the characteristic time.

The characteristic time for a molecular dynamics simulation may be estimated by considering an ordinary infra red spectrum. The peaks in an infra red spectrum correspond to the movements of a molecule. The characteristic time has to be substantially shorter than the fastest of these movements, and so it is necessary that it corresponds to a position way off the left hand end of the spectrum (Figure 5.7). The shortest period of oscillation in an infra red spectrum is about ten femtoseconds (about

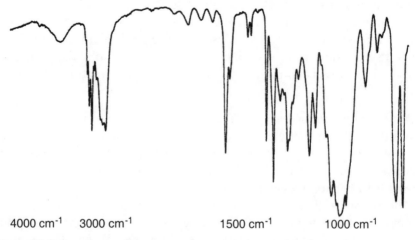

4000 cm⁻¹   3000 cm⁻¹   1500 cm⁻¹   1000 cm⁻¹

Figure 5.7 *Infra red spectrum*
(Supplied by Paul Wolstenholme-Hogg)

$3300\,\text{cm}^{-1}$). The time step usually chosen for molecular dynamics is one femtosecond ($33\,000\,\text{cm}^{-1}$).

This means that the molecular dynamics run calculates many structures for each oscillation of a carbon–hydrogen bond, and very many structures for the slower movements in the molecule. There are various tricks that can be used to speed things up. For example, the fastest thing that is happening in a molecular dynamics simulation is the vibration of bonds to hydrogen. It is possible to constrain the carbon–hydrogen bond lengths, on the grounds that this is unlikely to affect the movement of the structure very much. This turns out to be a reasonable approximation, and as a result the calculation runs faster, and it is possible to use a slightly larger time step, perhaps two femtoseconds instead of one (Ryckaert, 1985). It would be useful to constrain bond angles in the same way, but this does not work so well, because it seriously inhibits torsional rotations.

5.2.2.4 Length of a Simulation. How long should a molecular dynamics simulation run for? The short answer is, 'As long as possible!' Usually it would be desirable to run it for longer than is feasible, but the choice of a termination time will depend on the information required. A simulation might only last for about ten picoseconds. In terms of the infra red spectrum, this is only about $3.3\,\text{cm}^{-1}$ and not so far from the right hand end of the spectrum. However, it should be enough to get an impression of the sort of movement that corresponds to peaks in the fingerprint region of the spectrum. As an alternative illustration of how long this is, imagine the ten picoseconds spread along a line ten metres long. Each step in the simulation, one femtosecond or so, would then correspond to one millimetre on the line, and each step represents a complicated calculation. A simulation of a few picoseconds will take hours of computer time. The first molecular dynamics simulation of a macromolecule was carried out in 1977, by McCammon, Gelim and Karplus. This simulated the movement of bovine pancreatic trypsin inhibitor (Figure 5.8) over 8.8 picoseconds. Comparing this time with the infra red spectrum again, this is not so far off the right hand end (about $4\,\text{cm}^{-1}$). The protein was chosen because it is small (58 amino acids) and a good X-ray structure was available. It was not possible to include a detailed model of the solvent, but the four water molecules which were visible in the X-ray structure were included. The authors comment that the study has 'revealed a rich variety of motional phenomena ... at ordinary temperatures'. The protein was not rigid, but moved around at room temperature, without losing its secondary and tertiary structure. By 1990, computers had become so much more powerful that Karplus and

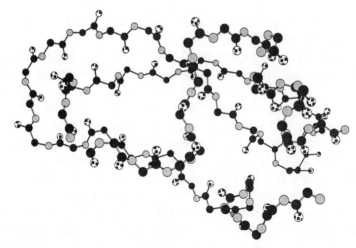

Figure 5.8 *Bovine pancreatic trypsin inhibitor*

Petsko could comment that for many problems 100 ps simulations were sufficient (Karplus and Petsko, 1990), and this is a reasonable length of time to run a simulation. However, even 100 ps is a very short time indeed.

The stages of a simulation. A molecular dynamics simulation takes a while to settle down. It begins with random components of energy, and these slowly distribute themselves around the molecule. At the beginning of a run, therefore, the energy tends to fluctuate over a large range, and then the range decreases as the run proceeds. The time required for the simulation to settle down depends on the molecule and on the choice of the starting structure. If the starting structure was very strained, it may be a long time before the molecule settles down, whereas if the starting structure was the only significant minimum of a structure then the simulation may produce consistent energies almost immediately.

A study of B-DNA (Figure 5.2) has been performed by Kollman, with this problem in mind (Rao and Kollman, 1990). A number of different starting structures for a twelve base pair sequence of DNA were used, and it was found that the starting structure had no effect on the results. The simulations ran for 84 ps, in this study, and the results suggest that 84 ps is sufficiently long to get consistent results in this case.

If the purpose of a study is just to find a few reasonable conformations, then there is no need to wait until the simulation overcomes its initial discomfort. More usually, however, if the energies of structures are to be compared, it is necessary to wait until the run has settled down, and then measure average values for the subsequent simulation (Figure 5.9).

How long is desirable? What about really difficult questions, such as the folding of proteins? In principle, if a molecular dynamics simulation

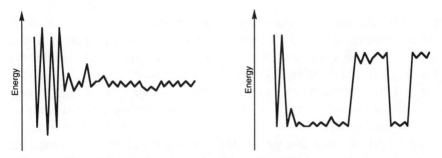

Figure 5.9 *The energy changes as a molecular dynamics simulation settle down, for a one state system (left) and a two state system (right)*

is run on an unfolded protein for long enough, it should fold up to its active form. Why has this not been done? Think of the ten metre line representing ten picoseconds again. Protein folding takes seconds to be complete, so the simulation would have to be run this long. The ten metre line represents a routine, but long, calculation. On the same scale, protein folding would require a line stretching from the earth to the sun. Computers are getting faster all the time, but there is no prospect they will get fast enough to do this sort of calculation in the foreseeable future (unless greatly simplified force-fields are used or many constraints can be introduced into the model: Skolnick and Kolinski, 1989).

5.2.3 Constant Energy or Constant Temperature?

In a molecular dynamics simulation, the total energy of the system can change from being mainly kinetic energy, in an unstrained conformation, to mainly potential energy, in a strained system. A ball bearing rolling in a potential well, such as one of those illustrated in Figure 5.5, would show this behaviour. At the bottom of the well, the ball bearing would move fast (high kinetic energy and low potential energy), and then slow down as it climbs the sides of the well, eventually stopping as it reverses direction. At this point, all of its energy is potential energy, and none is kinetic. The total energy of the ball bearing remains constant. This means that the ball bearing spends most of its time in high potential energy states, because these correspond to the time when it is moving slowly. This one-dimensional case is an extreme. In a multidimensional potential well, such as those describing molecules, it is unlikely that all of the energy will be concentrated as strain energy rather than kinetic energy. However, constant-energy molecular dynamics may not be an ideal description of molecular behaviour.

The temperature of a system corresponds to its kinetic energy, and not the sum of its kinetic and potential energy. An alternative approach is to use constant-temperature molecular dynamics instead of constant energy molecular dynamics. In the language of statistical mechanics, constant-temperature molecular dynamics should produce a canonical ensemble of states, whereas constant-energy molecular dynamics only produces a microcanonical ensemble. As the simulation proceeds, energy is added to the system when it cools down (*i.e.* the molecule is in a high energy conformation) and energy is removed from the system when it heats up (*i.e.* the molecule moves to a low energy conformation). The methods required to transfer energy to and from the system are sometimes referred to as stochastic dynamics (van Gunsteren and Berendsen, 1981).

Sometimes, a small part of a system is of particular interest, for example the active site of a large protein, or a molecule in a large amount of solvent. In such cases, a molecular dynamics simulation of the complete system may take a prohibitively long time, but it is possible to cheat by constraining the part of the system of lesser interest, and only allowing the interesting part to move. The difficulty with this approach lies in what to do at the interface between the moving and the non-moving parts, but stochastic dynamics can also be used to transfer energy in this situtation. Brooks and Karplus (1989) have used stochastic dynamics to investigate active sites, by constraining most of an enzyme and only allowing the active site to move. There was good agreement between this simulation and experimental data, and also with a conventional molecular dynamics simulation of the system. The conventional molecular dynamics simulation calculated the movement of 1531 atoms, while the stochastic dynamics approach only considered 294 explicit atoms, and average properties of the rest.

5.2.4 Simulated Annealing

As a method of finding the global minimum of a potential surface, molecular dynamics may be among the best options, using the process of simulated annealing, which was first described in a non-chemical paper in 1983 (Kirkpatrick *et al.*, 1983). Simulated annealing algorithms often work by making random changes to a structure, and then deciding if the new structure is better than the old, using the Metropolis criterion. Simulated annealing using molecular dynamics starts with a random configuration and random velocity for each atom, but then traces the movement of the molecule as the energy is slowly removed. It is very much more like real annealing than most simulated annealing algorithms, and it is guaranteed to work, provided the energy is removed infinitely slowly. A more rapid rate of cooling will remove the certainty of success, but this approach to finding global minima seems to work quite well. This sort of approach is often used to optimise structures generated by a distance geometry technique (see Chapter 4).

5.2.5 Structure Modification

If the crystal structure of a protein is available, and the question of whether there may be other accessible conformations for it arises, then molecular dynamics is the ideal tool. It can allow the protein to relax, and move around a little, without crossing the large energy barriers which other methods of conformational searching would not notice. A crystal structure may be used as a starting point to develop a range of sensible structures that a protein may actually adopt. Alternatively, n.O.e. data can be used as a starting point for a simulation. In this case there is unlikely to be enough data to define the structure uniquely, so a reasonable structure, consistent with the experiment, must be guessed, and this can be refined using molecular dynamics (Brünger and Karplus, 1991).

5.2.6 Free Energy Calculation

The differences between *energy*, *enthalpy* and *free energy* are often small, but the distinctions are important. The result of a force field calculation is ΔE_{MM}, and this will be approximately equal to the enthalpy change, ΔH, under many circumstances. The really interesting quantity, however, is the free energy change, ΔG, which is similar, but is also dependent on the entropy change, ΔS ($\Delta G = \Delta H - T \Delta S$). Equilibrium constants and rate constants are dependent on the free energy change, ΔG. In some circumstances, the difference between free energy and enthalpy is crucial. Free energy changes, ΔG, can be found by calculating the enthalpy change and the entropy change separately, and combining the results to give the free energy change, but the calculation of entropy is not very accurate.

The entropy of a system is related to its freedom to move about. Calculations of static conformations, by conformation searching as described in Chapter 4, do not give direct information about this. One way around this is to calculate the normal modes of oscillation of a conformation and to use these as a measure of the system's flexibility. (Go *et al.*, 1985) This is a time consuming procedure, and it depends on the assumption that the system is harmonic. This assumption is only reasonable for very small displacements from the minimum, and so the value obtained for the entropy is not very reliable.

A molecular dynamics simulation calculates how the molecule actually moves, and so it should be able to give a measure of entropy which takes account of the anharmonicity of the system. However, such results must be used with caution, as they only provide a good measure of the total entropy of the system if enough time has been devoted to the molecular dynamics simulation to be sure that the potential energy surface has been thoroughly explored.

5.2.6.1 Free Energy Perturbation. There is another way to find free energy changes using molecular dynamics, called free energy perturbation (Jorgensen, 1989). It can be shown that the change in free energy caused by a small alteration (or perturbation) to a system is equal to the enthalpy changes, averaged over many states,

$$\Delta G = -RT \ln \left\langle \exp\left(-\frac{H_{AB}}{RT}\right)\right\rangle_{A}$$

where $\langle \ \rangle_A$ is an ensemble average. The derivation and details of this relation need not concern us, (Zwanzig, 1954) but it has an important consequence. The free energy change for a process can be found without explicitly calculating the entropy, but by repeatedly calculating the enthalpy instead. This can only work if the change of interest is small, but that is not a serious drawback, because any large change can be split up into a series of small changes. The various states which form the average can conveniently be found using molecular dynamics.

To compare the free energy of two systems by this method, one system must be slowly changed into the another system; for example, a ligand can be introduced into a receptor site. This slow change does not correspond to any physical process, for atoms must be introduced a bit at a time, and charge must be changed slowly and continuously. Provided the start and the end of the process have physical significance, the pathway between them does not need to be physically possible.

The reliability of free energy perturbation calculations is difficult to determine. There have been excellent results reported, but the behaviour can depend sensitively on the shape of the potential surface being investigated. If the initial conditions are changed in a way that is small and reasonable, a result close to experiment can be changed to a result far from experiment (Wong, 1991; Straatsma and McCammon, 1992). The reason for the inaccuracy is probably that the changes between the initial and final states were made too rapidly. As with simulated annealing, a change which is infinitely slow would be best!

5.3 ANALYSIS OF RESULTS

5.3.1 The Difficulty of Analysis

It is very hard to analyse the results of a molecular dynamics run, because so much information is available. A typical ten picosecond run, using a two femtosecond time step, will generate five thousand intermediate structures, each of which will have energy and velocity information associated with them. If the molecule under consideration is large, this may be more than can be easily stored on a computer. Some way must be found to reduce this enormous quantity of information to a manageable and useful form.

5.3.2 Energy

The simplest way of analysing the results is just to find the average of the energy for all the structures. It is necessary to monitor how the enthalpy changes as the run progresses. It is likely that it will fluctuate widely at the beginning and then settle down to steadier fluctuations, and the average energy should just be taken from those structures after the wide fluctuations have ceased. This method of analysis is a little demoralising. A molecular dynamics run might take days or weeks of computer time, and at the end of this, you just have one number, the average energy, to show for it. It would be more satisfying to have more data available from the analysis.

5.3.3 Structure

It is possible to take averages of the structures as well as averages of the energies. This has an interesting effect. Methyl groups become straight lines, because they rotate, and so the average position of each hydrogen is on the axis of the bond to the methyl group carbon. The hydrogens of hydroxy groups fall onto the oxygen. This is illustrated in Figure 5.10. This sort of average may give some idea of how things behave. It is probably more meaningful to compile an average of particular parameters, for example a hydrogen bond length, or a hydrophobic interaction. This would not need very much storage space, because the average could be updated every step, and so the entire list of conformations would not need to be kept.

In a similar way, conformations could be selected every hundred femtoseconds or so, and these compared, or superimposed. This would be better than considering every conformation, because adjacent conformations are likely to be very similar, and there will be a limit to the number of structures it is possible to superimpose and view. Molecular dynamics can also be used to generate a series of conformations, just like a conformational search method described before. The different conformations are taken from the complete molecular dynamics sequence, and each is minimised. This will give a series of minima, which can be analysed by a Boltzmann distribution.

Figure 5.10 *An average structure from a simulation of D-alanine-D-alanine bound to aglyco-ristocetin*

5.3.4 Movies

One way to express the results of a molecular dynamics simulation is to show the structures generated as a 'movie'. This means that the consecutive steps of the simulation are rapidly displayed one after another, to give an impression of movement. Of course, this takes a great deal of memory, and 'movies' produced in this way tend to be only a few seconds long, but it gives a good feeling for the way molecules writhe around. Once you have seen methyl groups spinning, side chains twisting and turning, hydrogen bonds breaking and reforming, and the other complex aspects of molecular movement, static CPK models will never look the same again.

This is a very effective method of presenting results qualitatively, but it requires a large amount of computer time and disk space, and it can provide no quantitative information. It is probably not the most useful method of analysing molecular dynamics simulations.

5.4 KEY POINTS

- *Molecules are not static but are in continuous movement.* A molecular dynamics simulation is a calculation of how molecules move, and so it is closer to reality than other methods of extracting information from force fields.

- Molecular dynamics investigates the low lying regions of the potential surface, not just the minima, because a molecule will be able to cross low potential energy barriers at normal temperatures.
- Molecular dynamics can be used to limit searches to accessible conformations, rather than the larger number of possible conformations, because the molecule will not be able to cross large energy barriers. Conformations which are of low energy but not accessible, because a large barrier would have to be crossed in order to reach them, will not be covered by the simulation
- Molecular dynamics can be used to find entropy, S, and free energy, G, not just the internal energy, U.
- These advantages mean it is particularly suited to finding accessible protein conformations from crystal structures, and for estimating binding constants, when solvent effects are important and comprehensive searches of the potential energy surface may be unrealistically large. However, its complexity prevents it being the method of choice for simpler problems, such as the conformational analysis of small molecules *in vacuo*.

QUESTIONS

Q5.1 Simple alkanes

How long a simulation is required to explore the conformation space of butane?

Q5.2 Methylcyclohexane ring flipping

How do the structure and energy of methylcyclohexane (Figure 5.11) change as a simulation proceeds? Does the ring flip from equatorial to axial? If the simulation starts with axial methylcyclohexane does this make any difference? Do the molecules behave as you would have predicted?

equatorial   axial

Figure 5.11

Q5.3 Hydroxyketone dynamics

A conformation search on the structure shown in Figure 5.12 will probably show that the global minimum structure has a hydrogen bond between the alcohol and the ketone.

How does the hydrogen bond length differ in the global minimum and during a molecular dynamics simulation? What will happen if a solvent model for water is used?

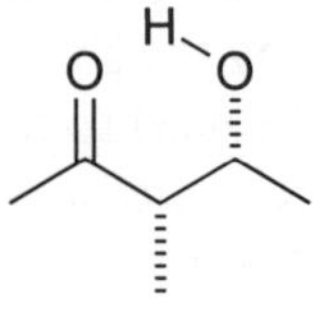

Figure 5.12

Q5.4 Cram's Rule

Can molecular dynamics be used to analyse the conformational preferences of the molecules shown in Figure 5.13 and so predict the preferred face for nucleophilic attack in each case? This question was answered by a conformation search in Chapter 4. How does molecular dynamics compare?

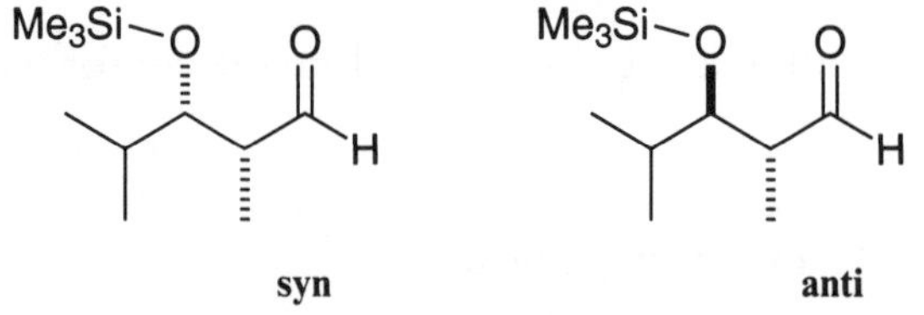

Figure 5.13

Q5.5 Links and knots

What is the smallest cycloalkane which can form a catenane (two linked rings)? What is the smallest cycloalkane which can comfortably form a knot?

Q5.6 Ristocetin

How do ligands bind to aglyco-ristocetin (Figure 5.14)? This is one of the simplest possible models of ligands binding to proteins.

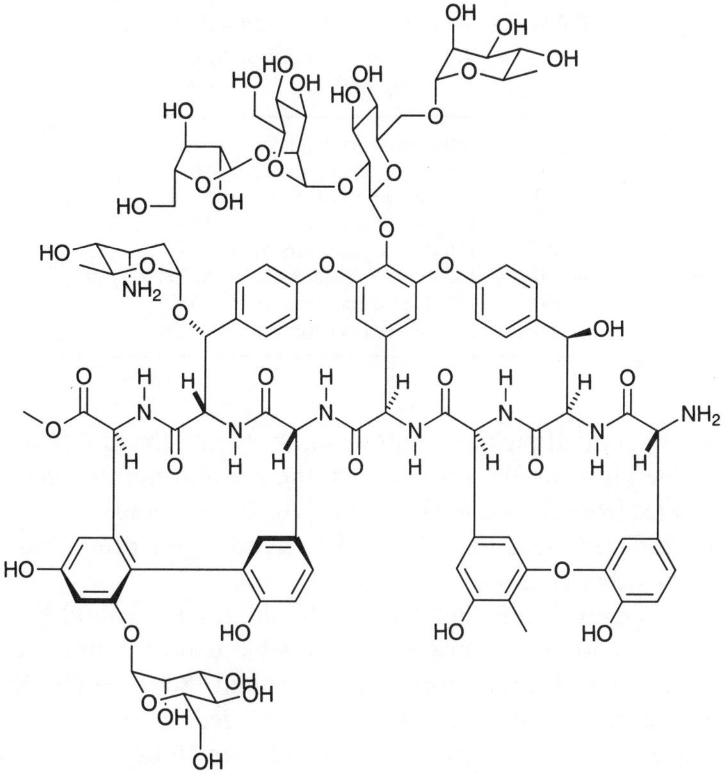

Figure 5.14 *Ristocetin*

DISCUSSION

D5.1 Simple alkanes

Several molecular dynamics simulations of butane were run using MacroModel and MM2*, whilst monitoring the C—C—C—C torsion angle. The results of the first are given in Table 5.1. The simulation was run for 1 ns, which is quite a long time, by the standards of molecular dynamics simulations. Every 15 fs (ten steps), the C—C—C—C angle was measured. The results are given in the table.

Even this very long simulation did not give the butane an opportunity to move from the extended conformation. This is not the result which might have been expected. The simulation was repeated four times, this time using the staggered conformation (C—C—C—C = 60°) as the starting structure, and then run once for 10 ns instead of 1 ns. The results are given in Table 5.2. The first two runs are identical. This is because the energy is initially distributed randomly around the molecule, and this distribution determines the course of the simulation. Computers are not

Table 5.1 *1 ns simulation of butane, 300 K, 1.5 fs steps, starting with C—C—C—C at 180°*

140 – 150 degree occurrences =	29
150 – 160 degree occurrences =	5644
160 – 170 degree occurrences =	15973
170 – 180 degree occurrences =	11544
−180 – −170 degree occurrences =	11653
−170 – −160 degree occurrences =	16229
−160 – −150 degree occurrences =	5557
−150 – −140 degree occurrences =	37

always very good at picking random numbers, and the first two columns of the table illustrate the results when the random number generator is started twice from the same place. It is possible to avoid this problem by choosing different 'seeds' for the random number generator, and this was done in the other runs.

It might be expected that butane should rotate around the central C—C bond, and so explore all the possible conformations, although spending most of its time with the angle around 180° or ±60°. Since this is not the result in Table 5.2, it may be concluded that these runs are not long enough! If even butane requires more than 10 ns of simulated time to explore its conformation space, what hope is there for more complex structures?

Fortunately, butane is a particularly difficult case. Larger molecules will have more energy spread throughout all of their atoms, but there is a reasonable chance that enough of it will concentrate in a particular torsion angle, from time to time, to allow the torsion angle to overcome its rotational barrier. With butane, there are so few degrees of freedom that the chances of all of the energy being concentrated in just one of them is rather small. If this reasoning is correct, then we should expect that longer alkanes will explore their conformation spaces more rapidly. The results for such simulations are given in Tables 5.3 and 5.4.

This effect will be most important *in vacuo*. Butane in solution will be able to exchange energy with the solvent, and so would explore its conformation space more rapidly.

Table 5.3 shows the results of one 10 ns simulation of hexane, and shows the results of monitoring all three C—C—C—C torsion angles. The two terminal angles now give the pattern that might be expected, covering all possible values of the torsion angles and concentrating in the areas of lower energy. The central torsion angle has also explored all low energy orientations, but has spent less time between 0° and 120° than might have been expected. Presumably this is because some orientations

Table 5.2 *Molecular dynamics simulation (1 ns and 10 ns) on butane with C—C—C—C = 60° as the starting structure (300 K)*

	1 ns	1 ns	1 ns	1 ns	10 ns
−180 − −170 degree occurrences	8707	8707	5070	2	112969
−170 − −160 degree occurrences	11853	11853	7490	2	144449
−160 − −150 degree occurrences	3716	3716	3022	2	49060
−150 − −140 degree occurrences	64	64	174	4	728
−140 − −130 degree occurrences	11	11	58	4	6
−130 − −120 degree occurrences	7	7	11	7	11
−120 − −110 degree occurrences	24	24	92	48	44
−110 − −100 degree occurrences	157	157	460	306	284
−100 − −90 degree occurrences	938	938	1669	1939	2300
−90 − −80 degree occurrences	2906	2906	5104	6500	7492
−80 − −70 degree occurrences	4244	4244	7824	9899	12807
−70 − −60 degree occurrences	3986	3986	8396	10226	13038
−60 − −50 degree occurrences	3521	3521	6502	8144	9696
−50 − −40 degree occurrences	1835	1835	3646	4306	5178
−40 − −30 degree occurrences	401	401	888	891	1051
−30 − −20 degree occurrences	30	30	137	69	92
−20 − −10 degree occurrences	0	0	5	8	4
10 − 20 degree occurrences	0	0	0	1	0
20 − 30 degree occurrences	0	0	15	34	0
30 − 40 degree occurrences	0	0	26	389	0
40 − 50 degree occurrences	0	0	29	2450	0
50 − 60 degree occurrences	0	0	31	5312	0
60 − 70 degree occurrences	0	0	23	5456	0
70 − 80 degree occurrences	0	0	39	6199	0
80 − 90 degree occurrences	0	0	32	3573	0
90 − 100 degree occurrences	0	0	32	754	0
100 − 110 degree occurrences	0	0	19	95	0
110 − 120 degree occurrences	0	0	22	23	0
120 − 130 degree occurrences	4	4	15	8	0
130 − 140 degree occurrences	14	14	54	4	16
140 − 150 degree occurrences	97	97	154	4	654
150 − 160 degree occurrences	3961	3961	3180	2	49306
160 − 170 degree occurrences	11401	11401	7255	3	144364
170 − 180 degree occurrences	8789	8789	5192	2	113117

of the terminal groups make twisting the central torsion angle much more difficult.

Table 5.4 shows the result of a similar 10 ns simulation of decane. In this simulation, all of the torsion angles appear to have explored all of the possible orientations in a similar sort of way. It seems, therefore, that in some respects, molecular dynamics gets easier as the molecules being studied get larger!

Stochastic dynamics provides an alternative way of making the system larger, by adding a heat bath, without greatly increasing the computational cost of the simulation. A one nanosecond simulation of butane

Table 5.3 *Molecular dynamics simulation (10 ns) on hexane (300 K)*

Torsion angles:		1–2–3–4	2–3–4–5	3–4–5–6
−180 – −170°	=	165633	155350	158793
−170 – −160°	=	99295	110715	95428
−160 – −150°	=	33236	38151	33843
−150 – −140°	=	5825	5155	6226
−140 – −130°	=	696	553	657
−130 – −120°	=	91	211	131
−120 – −110°	=	125	253	128
−110 – −100°	=	408	507	424
−100 – −90°	=	1396	1651	1551
−90 – −80°	=	3846	5160	5454
−80 – −70°	=	7425	10438	11988
−70 – −60°	=	8228	12703	15622
−60 – −50°	=	5386	9454	11321
−50 – −40°	=	2310	3943	4402
−40 – −30°	=	619	873	933
−30 – −20°	=	91	88	144
−20 – −10°	=	22	0	41
−10 – 0°	=	0	0	4
0 – 10°	=	0	0	22
10 – 20°	=	14	0	18
20 – 30°	=	156	6	90
30 – 40°	=	785	7	483
40 – 50°	=	3057	5	1995
50 – 60°	=	7008	20	5119
60 – 70°	=	9855	59	6891
70 – 80°	=	9143	35	6177
80 – 90°	=	5315	11	3424
90 – 100°	=	2106	27	1273
100 – 110°	=	668	11	367
110 – 120°	=	227	7	129
120 – 130°	=	204	55	143
130 – 140°	=	736	804	710
140 – 150°	=	5597	5837	5731
150 – 160°	=	31790	39117	31251
160 – 170°	=	94360	112036	94984
170 – 180°	=	161013	153424	160769

gives a good distribution of conformations, because the molecule is given extra energy as barriers are crossed (Table 5.5).

D5.2 Methylcyclohexane ring flipping

A molecular dynamics simulation of methylcyclohexane (Figure 5.11) makes the structures move around in an entertaining way, and watching a few hundred femtoseconds of the simulation is quite interesting.

Table 5.4 *Molecular dynamics simulation (10 ns) on decane (300 K)*

Torsion angles:		1–2–3–4	2–3–4–5	3–4–5–6	4–5–6–7	5–6–7–8	6–7–8–9	7–8–9–10
−180 –	−170° =	139850	126247	146272	142220	140865	152708	130427
−170 –	−160° =	81130	71100	85082	82146	81903	85928	75190
−160 –	−150° =	28283	24558	30150	29352	28713	29279	25842
−150 –	−140° =	6991	6638	7538	7743	7385	7642	6029
−140 –	−130° =	1455	1604	1679	1907	1933	1991	1308
−130 –	−120° =	526	503	389	607	716	680	520
−120 –	−110° =	647	742	278	615	616	734	571
−110 –	−100° =	1317	1663	610	1182	1343	1412	1449
−100 –	−90° =	3583	4535	2038	2412	4181	3924	4406
−90 –	−80° =	10997	12661	5171	6205	10835	10263	11538
−80 –	−70° =	20892	23451	9364	10559	20253	18175	22164
−70 –	−60° =	23870	25980	11224	12188	22371	21320	25678
−60 –	−50° =	16699	17114	7616	8434	14755	14867	17699
−50 –	−40° =	6879	6845	3097	3407	5867	5720	7547
−40 –	−30° =	1827	1735	666	882	1523	1512	1879
−30 –	−20° =	362	313	101	140	227	260	341
−20 –	−10° =	77	68	7	29	40	59	68
−10 –	0° =	19	29	0	2	15	11	16
0 –	10° =	11	19	1	4	4	14	9
10 –	20° =	39	101	76	30	14	35	65
20 –	30° =	309	372	318	234	175	183	301
30 –	40° =	1497	2083	1810	1651	1096	666	2125
40 –	50° =	4905	8368	6111	6378	4363	2613	7237
50 –	60° =	10389	19964	15015	16333	10562	5949	18817
60 –	70° =	15485	29796	20907	24543	15748	8177	28719
70 –	80° =	13872	26169	18614	21892	14551	7821	25455
80 –	90° =	7509	14767	9977	11288	8310	4849	12780
90 –	100° =	2921	5394	3865	4100	3362	2051	4490
100 –	110° =	1126	1442	1278	1285	1476	1035	1576
110 –	120° =	608	627	864	594	651	605	612
120 –	130° =	524	481	776	741	702	613	632
130 –	140° =	1589	1452	2040	2113	1448	1825	1457
140 –	150° =	6845	6060	8992	7332	7020	7304	6044
150 –	160° =	28952	25323	32358	29327	30068	29356	24022
160 –	170° =	82737	72628	86664	84515	82736	85061	71154
170 –	180° =	141944	125834	145718	144276	140839	152024	128499

However, the methyl groups will not flip from axial to equatorial, or *vice versa*, in a few picoseconds, and this is the maximum length of 'movie' that most people are prepared to sit through. The rings will flip eventually, but it takes a very long time. It would be possible to carry out a simulation for milliseconds (millions of millions of steps!) and then edit the exciting part in which the rings flip, but the results would not be a representative picture of the behaviour of this system.

Heating the system makes the rings flip more often. However, putting more energy into the system makes the bonds stretch alarmingly, using

Table 5.5 *Stochastic dynamics simulation (1 ns) on butane (300 K)*

−180 − −170 degree occurrences =	12968
−170 − −160 degree occurrences =	6602
−160 − −150 degree occurrences =	2068
−150 − −140 degree occurrences =	525
−140 − −130 degree occurrences =	138
−130 − −120 degree occurrences =	48
−120 − −110 degree occurrences =	51
−110 − −100 degree occurrences =	143
−100 − −90 degree occurrences =	582
−90 − −80 degree occurrences =	1853
−80 − −70 degree occurrences =	3813
−70 − −60 degree occurrences =	4606
−60 − −50 degree occurrences =	3188
−50 − −40 degree occurrences =	1199
−40 − −30 degree occurrences =	278
−30 − −20 degree occurrences =	23
0 − 10 degree occurrences =	1
10 − 20 degree occurrences =	17
20 − 30 degree occurrences =	45
30 − 40 degree occurrences =	169
40 − 50 degree occurrences =	545
50 − 60 degree occurrences =	1390
60 − 70 degree occurrences =	1747
70 − 80 degree occurrences =	1396
80 − 90 degree occurrences =	605
90 − 100 degree occurrences =	182
100 − 110 degree occurrences =	76
110 − 120 degree occurrences =	41
120 − 130 degree occurrences =	35
130 − 140 degree occurrences =	96
140 − 150 degree occurrences =	500
150 − 160 degree occurrences =	2158
160 − 170 degree occurrences =	6788
170 − 180 degree occurrences =	12790

most standard force fields, and the force field is no longer a good model of the structure. Fairly modest temperature, such as 1000 K, can make the bond lengths change by an unrealistic amount.

The results of four 10 ns simulations of methylcyclohexane at elevated temperatures are given in Table 5.6. One hundred structures were sampled from each run and these were minimised. On minimisation, all of the structures became one of the five conformations shown in Figure 4.22. The two lowest energy structures are chairs and the rest are boats. Two of the runs used SHAKE, a routine which allows larger time steps to be taken by constraining the C—H bonds (Ryckaert, 1985).

The table shows that the boat form is favoured at high temperatures, to a greater extent than would be expected by calculating the Boltzmann

Table 5.6 *Methylcyclohexane dynamics (10 ns) at 1000 K and 10 000 K*

Conformation[a]	1000 K SHAKE	1000 K	10 000 K SHAKE	10 000 K
1	65	50	23	17
2	18	24	17	13
3	8	15	25	28
4	8	8	19	26
5	2	4	14	15
C—H bond length	1.1 Å	0.95–1.30 Å	1.1 Å	0.77–1.63 Å
C—C bond length	1.4–1.7 Å	1.42–1.75 Å	1.20–2.15 Å	1.16–2.10 Å
Amount of time in boat conformation	18%	24%	59%	70%

[a] The conformations are those given in Figure 4.22.

factors of the minima at these temperatures. This suggests that the entropy of the boat form is higher than that of the chair, and so the boats are increasingly favoured as the temperature increases. The table also gives the ranges of bond lengths which are seen throughout the simulation. Even at 1000 K, the bonds are stretching more than is physically reasonable. Raising the temperature of the simulation will, therefore, make conformation changes occur more rapidly, but will move the simulation away from those regions of conformation space which the force field can describe well.

D5.3 Hydroxyketone dynamics

The average distance between the carbonyl oxygen and the hydroxyl hydrogen (Figure 5.12) is 2.0 Å during a 100 ps simulation at 300 K, using the AMBER force field and implicit hydrogen atoms. Using a continuum water model, this distance increases to 4.7 Å, as the strength of the intramolecular hydrogen bond is diminished and the hydrophobicity of the methyl groups becomes significant.

A Monte Carlo conformations search could be performed on the same system, and the average O⋯H distance can be estimated by the averaging the measurements from different minima, weighted by their Boltzmann factors. The result of this procedure is not precisely the same as the molecular dynamics average, and in this case molecular dynamics is more likely to give a result close to experimental values.

D5.4 Cram's Rule

A molecular dynamics simulation of the system shown in Figure 5.13 can give values for how the dihedral angle Me—C—C=O varies with

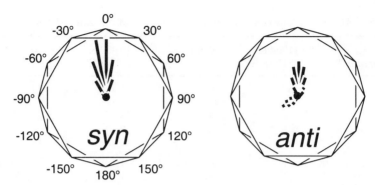

Figure 5.15 *Variation of Me—C—C=O angle with time*

time, and the results are illustrated in Figure 5.15. The results are consistent with those from the Monte Carlo analysis of the same system. The *syn* isomer has a strong preference for the methyl group to eclipse the carbonyl group, which will lead to *re* face attack. The *anti* isomer spends some of its time with the methyl group eclipsing the carbonyl and some with the methyl group at 120° to the carbonyl, which would cause *si* face attack. It is immediately obvious from the figure that the selectivity is likely to be higher in the case of the *syn* compound. The preferred face of attack for the *anti* isomer is less clear, which is consistent with the lower selectivity which is observed for this isomer.

D5.5 Links and knots

The energy required to break a C—C bond is about $360 \, kJ \, mol^{-1}$. It will not be possible to form a catenane if the energy required to link two rings together is greater than this. It is unlikely to be possible to form a catenane unless the energy required to link two rings is very much less than this. A Monte Carlo conformation search will find the energy of a single ring very effectively, but will work far less well for linked rings, because it will be hard to constrain the search so that the rings stay linked. With molecular dynamics, this is not a problem, provided that the temperature is kept low enough so that the simulation stays physically significant. Using MM2*, it is possible to calculate that a $C_{20}H_{40}$ cycloalkane will have a global minimum energy of $84.1 \, kJ \, mol^{-1}$ or less (the ring is so large that it is hard to be confident that the global minimum has really been found). A molecular dynamics simulation was run on two linked $C_{20}H_{40}$ rings (Figure 5.16), and series of structures were taken from the simulation. All of the structures were linked rings, and minimisation gave a structure with an energy of

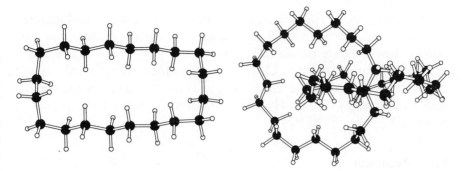

Figure 5.16 *Unlinked and linked $C_{20}H_{40}$*

$339.3 \, \text{kJ mol}^{-1}$. This is only $171 \, \text{kJ mol}^{-1}$ higher in energy than the unlinked rings, so it is just possible to imagine making such a structure. It is likely that larger ring sizes are needed to reduce the energetic penalty of linking the rings.

Similar considerations apply when considering the smallest cycloalkane which could form a knot. Molecular dynamics simulations of knotted rings gave very high energies of rings containing less than sixty carbon atoms, but reasonable energies for larger rings. The conformation of $C_{60}H_{120}$ illustrated (Figure 5.17) has an energy of $164 \, \text{kJ mol}^{-1}$, compared with an energy of $153 \, \text{kJ mol}^{-1}$ which has been found for an

Figure 5.17 *$C_{60}H_{120}$*

open $C_{60}H_{120}$ loop. These energies are unlikely to be the global minima, but the molecular dynamics simulation has demonstrated that low energy structures are possible for rings of this size. Frisch and Wasserman (1961) considered that $C_{50}H_{100}$ may be able to form a knot, based on mechanical models, but MM2* suggests that this may be too small a ring, as the energy of a $C_{54}H_{108}$ knot appears to be about $300\,kJ\,mol^{-1}$, and $C_{48}H_{96}$ over $500\,kJ\,mol^{-1}$.

D5.6 Ristocetin

Molecular dynamics is a good way of calculating binding constants. For example, the free energy of association of 18-crown-6 with a potassium ion in water (Figure 5.18) has been calculated (Dang and Kollman, 1990). Nine hundred and seven water molecules were included in the model to solvate the system. This sounds like a great many, but it only corresponds to a layer about 4 molecules thick around the system. A molecular dynamics simulation was run for 500 ps, and the calculated free energy of association is $-8\,kJ\,mol^{-1}$ which compares well with the experimental value of $-12\,kJ\,mol^{-1}$.

Ristocetin bound to Ac-D-Ala-D-Ala is a much larger system (Figure 5.19), and so the calculation will be much harder, particularly if allowance is made for the dimerization of the complex. A study (Groves *et al.*, 1995) used a continuum solvent model (Still *et al.*, 1990) instead of explicit water molecules, in order to simplify the calculations. The simulation demonstrated that the experimentally measured nOes were consistent with the molecular mechanics model, which may be seen as increasing confidence in the model. A calculation of the binding energy would require a very long simulation of both the bound and the free states of the system, and the continuum solvent model will introduce a systematic error between these two states. An easier calculation would be to compare the binding energies of the different diastereoisomers of the peptide ligand, as this would not require the comparison of the unbound energy with the bound energy. This may be useful in designing new ligands for this or similar systems. Alternatively, it would also be

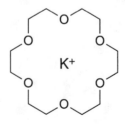

Figure 5.18 *18-crown-6 and a potassium ion. 907 water molecules are omitted for clarity*

Figure 5.19 *Monomeric aglyco-ristocetin bound to acetyl-D-Ala-D-Ala*

reasonable to compare the binding of the ligand with analogues of ristocetin, and so, perhaps, design other molecules with similar or greater binding affinities.

CHAPTER 6

Molecular Orbital Theory

'The underlying physical laws necessary for the mathematical theory of a large part of physics and the whole of chemistry are thus completely known, and the difficulty is only that the exact application of these laws leads to equations much too complicated to be soluble.'

P. A. M. Dirac

6.1 WHY USE QUANTUM MECHANICS?

Since the introduction of quantum theory in the early part of the twentieth century and the realisation that particles can behave like waves as well as particles, the application of quantum mechanics to the theory of atomic and molecular structure has been extremely successful. With seventy years of experience, it is reasonable to assume that the theory is very reliable and can describe chemistry, just as Dirac said (Dirac, 1929). Why, then, bother with molecular mechanics at all? The problem is that the equations which describe molecular structure are still too complex to solve precisely for anything except the smallest of systems. As a result, a wide range of approximations have been developed to extend these methods to larger molecules, which is to say, those containing more than one atom. Even with the most powerful computers available today, the sort of calculations which have been discussed in the earlier chapters are beyond the reach of quantum mechanics. A minimisation of a biologically active molecule with a molecular weight of one thousand or so may be possible, but the thousands of calculations required for a conformation search or a molecular dynamics simulation are unlikely to be feasible.

6.2 ATOMIC ORBITALS

Rutherford knew that atoms have tiny nuclei, and he assumed that the electrons were in stable orbits around them, rather like planets. This was found to be impossible, because accelerating electrons (this includes electrons moving in a circular orbit) give out photons, and so atoms should collapse in a flash of light, unless there is something holding the electrons in their orbits. Bohr's model of atomic structure assumes this does not happen, but does not give an explanation for atomic stability. Prince Louis de Broglie realised that particles can be considered as waves, and this solves the problem, because electrons can form 'standing waves' around the atomic nucleus. The energy of the standing wave depends on the number of nodes it contains (The same is true of skipping ropes. A rope spun in the usual way has a node, that is, is stationary, at each end, where it is held. It is possible to spin a rope so that it has another node in the middle, but this requires more vigorous spinning). Using the wave equation developed by Erwin Schrödinger, a complete description was developed of isolated, non-relativistic one-electron atoms, which are rather uncommon. Fortunately, relativistic effects in hydrogen atoms are very small, so the theory could be used to calculate the properties of hydrogen atoms very precisely.

6.2.1 One-electron Atoms

This approach predicts that electrons exist in *orbitals* of particular energies. The properties of these orbitals may also be found from the Schrödinger equation in terms of a wavefunction, Ψ. It is postulated that the square of the wavefunction corresponds to the probability of finding an electron in a particular place. Unlike probability, the wavefunction can be negative (or even complex), and this has important consequences for the way orbitals interact with each other.

The orbitals are numbered in order of energy, by the *principal quantum number*, n. The lowest energy orbital corresponds to $n = 1$, the next lowest $n = 2$ and so on. The lowest energy corresponds to just one orbital, which is spherical, but all the higher energies correspond to several orbitals. The second lowest energy corresponds to four orbitals, usually labelled 2s, $2p_x$, $2p_y$, $2p_z$. (For multi-electron atoms, the 2s orbital has a lower energy than the 2p orbitals.) The letters correspond to the *azimuthal quantum number*, l, which takes values from zero to $n - 1$. Therefore, when $n = 1$, l must be zero, and a value of zero is labelled 's' for 'sharp', a description which relates to lines in atomic spectra. Thus there is a 1s orbital, but no 1p orbital. When $n = 2$, then l can be zero (s) or one, labelled 'p' for 'principal', and so there are 2s and 2p orbitals. The series continues $l = 2$ [d (diffuse)], $l = 3$ [f (fundamental)], $l = 4$ (g), and so on. The azimuthal quantum number also reflects the symmetry of the orbitals: s orbitals are spherical, p orbitals

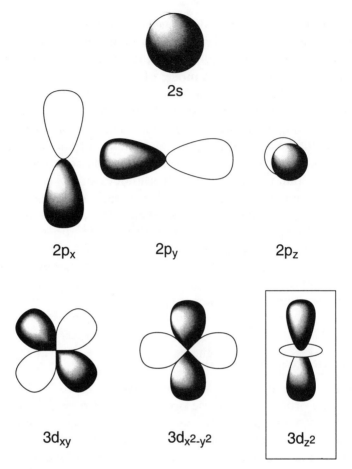

Figure 6.1 *Some atomic orbitals. The $3d_{z^2}$ orbital is shown perpendicular to the others to better illustrate its shape*

are dumbbell shaped, d orbitals resemble a four-leaved clover. Some orbitals are illustrated in Figure 6.1 and Plate 5.

The principal quantum number, n, also corresponds to one more than the number of nodes in the orbital. A node, in this context, means a stationary point in a standing wave, and so corresponds to a region where the wavefunction is zero. A 1s orbital, therefore, has no nodes, and the wavefunction is non-zero everywhere. A $2p_x$ orbital must have one node, and this is clear in Figure 6.1. The $2p_x$ orbital has a horizontal nodal plane. The $3d_{xy}$ orbital must have two nodes, and these correspond to a vertical and a horizontal plane. The node in the 2s orbital is not obvious from the diagram. This orbital has a radial node. At a certain distance from the nucleus, the wavefunction is zero, and the sign of the wavefunction changes.

If all three p orbitals are added together, the corresponding electron densities combine to form a sphere. This may seem surprising, since p orbitals are usually drawn as dumbbells, but the sharp edge to the dumbbell is a convenient way of depicting the orbital, rather than a precise description of it nature. Orbitals really fade away, and do not stop suddenly. In the same way, the d_{z^2} orbital may be regarded as the sum of a $d_{z^2-y^2}$ and a $d_{z^2-x^2}$ orbital.

Atomic orbitals are *orthogonal* to each other. This means that if the values of the wavefunctions for two different orbitals on the same atom are multiplied and added up, the result will be zero. This is clear for the product of a 2s orbital and a $2p_x$ orbital, because the two halves of the $2p_x$ orbital have opposite signs, but equal size. In the same way, the product 2s orbital with a $3d_{xy}$ orbital will have two positive areas and two negative areas, so the sum will be zero. This is not so obvious for a 1s orbital interacting with a 2s orbital, but it is also true, because the 2s orbital has a radial node.

The number of orbitals of each type is $2l + 1$, and they are labelled by the *magnetic quantum number, m*. This has values $l, l-1, \ldots, 0, \ldots, -l$. Electrons have *spin*, a quantum mechanical property (*spin quantum number, s* $= \pm\frac{1}{2}$) that, in some respects, is more like charge than a macroscopic idea of spinning.

6.2.2 Multi-electron Atoms

For atoms with more than one electron the situation becomes much more complicated, and it is impossible to calculate exactly what happens. If it is assumed that the electrons are in orbitals like those for one electron atoms, and also assumed that each electron is only affected by the average of all the others, then it is possible to calculate what happens. The second assumption is particularly perilous. Imagine crossing a road, assuming an average traffic distribution. Whenever you cross, you will be partly run over, so it makes no difference if you look left and right or if you close your eyes and run. However, these assumptions turn out to work quite well for molecules.

One result of having several electrons in the atom is that the s orbitals become lower in energy than the corresponding p orbitals, which are lower than the corresponding d orbitals. Two electrons cannot have the same quantum numbers (this is called the Pauli Principle), so only two electrons fit in each orbital, and they must have opposite spins.

6.3 MOLECULAR ORBITALS

It is impossible to do calculations on multi-electron atoms without approximations. Molecules are even harder, and so more approximations must be introduced. However, it is possible to build a qualitative picture of molecular orbitals fairly easily, by assuming that molecular orbitals are formed from *linear combinations of atomic orbitals*. This means that a molecular orbital is made up of the bits of the atomic orbitals on each atom that seem to be pointing in about the right direction.

6.3.1 Orbital Interactions

Atomic orbitals on one atom are *orthogonal* to each other, and do not interact. Atomic orbitals on different atoms are not necessarily orthogonal, and so they can interact (Figure 6.2). Atomic orbitals on different atoms may interact to form *molecular orbitals*. Whatever happens, however, the total number of orbitals is conserved.

6.3.1.1 Interaction of Two Atomic Orbitals.   Provided the symmetry of the orbitals allows a net interaction, two atomic orbitals interact to form two molecular orbitals, one of higher energy than both the atomic orbitals, and the other of lower energy than both the atomic orbitals (Figure 6.3). The interaction is strong if the atomic orbitals are of similar energy to each other, and if there is good overlap, which means the orbitals must be close to each other and of similar sizes. Molecular orbitals are labelled by their symmetry, like atomic orbitals, but the label corresponds to the symmetry around the axis of the interacting centres. In order to distinguish them from atomic orbitals, molecular orbitals are labelled with Greek letters: σ instead of s, π instead of p, and so on.

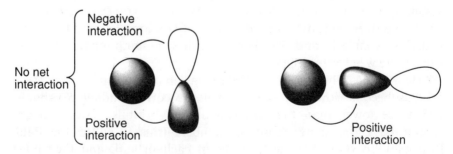

Figure 6.2 *Interaction of atomic orbitals*

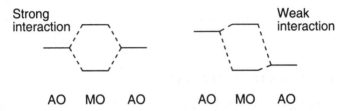

Figure 6.3 *Interactions diminish as atomic orbitals become less close in energy*

6.3.1.2 Interaction of Several Atomic Orbitals. If more than two atomic orbitals overlap, then all of them will interact with all of the others, and it is hard to work out what is happening. The problem can be approached by deciding which of the atomic orbitals will interact most strongly, and pairing them in order of strength of interaction. The multiple interactions can then be considered, to a first approximation, as a series of pairwise interactions, and an intermediate molecular orbital diagram constructed, which allows only for the strongest pairwise interactions. After the strongest interactions have been considered, the remaining interactions can be introduced. These must be smaller, and so can be treated as perturbations on the intermediate diagram.

6.3.1.3 Hybridisation. The molecular orbital description of bonding can be used to describe many puzzling phenomena, but it is rather a long way from the classical model, in which bonds are composed of pairs of electrons. Molecular orbitals are spread through the whole molecule, and it is not easy to see how they can be related to two-centre, two-electron bonds. One way to do this is to use the concept of hybridisation.

Atomic orbitals do not have geometries that look as if they can form bonds in the expected way. They can be combined to give orbitals that point along the conventional bonding directions, as a stepping stone on the way to describing the complete system. One s orbital and three p orbitals can be combined to give four equivalent sp^3 orbitals. The total energy of the four sp^3 orbitals is the same as the total energy as the one s and three p orbitals, and the sp^3 orbitals only exist as a stepping stone to considering the molecular orbitals of the whole system. The advantage of hybridisation is that the new orbitals point along bonds, so it is much easier to see how they interact with hybrid orbitals on neighbouring atoms.

In many cases, hybridisation is a helpful model of atomic structure, but it cannot cope with all systems. In particular, it does not give a reasonable picture of the bonding in oxygen, which is a diradical, or in diborane, which contains bridging hydrogens. Molecular orbital models without hybridisation can cope with both these molecules.

6.4 USING QUANTUM MECHANICS

Quantum mechanics is hard because normal macroscopic concepts do not always apply. Schrödinger's equation gives a wave function, the square of which corresponds to a probability of finding an electron at a particular point. This shows that orbitals do not have clearly defined edges which contain their electron, but the probability of finding an electron diminishes gradually with distance. The orbitals sketched in Figures 6.1 and 6.2 are misleading because they do not illustrate this, but it is a very difficult concept to draw clearly. The chance of finding an electron within the orbitals as drawn may be less than 80%. However, the probability density is very low at the surface which is illustrated, and becomes lower further away from the centre of the orbital (Plate 5).

6.4.1 Simple Molecular Orbital Calculations

What do molecular orbitals look like? One way of calculating their structures is to assume that molecular orbitals can be expressed as linear combinations of atomic orbitals (LCAO), as shown in Equation 6.1:

$$\Psi = c_1\phi_1 + c_2\phi_2 + \ldots + c_n\phi_n \tag{6.1}$$

where ϕ_i are the atomic orbitals; Ψ is a molecular orbital; c_i are coefficients.

The energy of an orbital can be found from its wavefunction, by use of Schrödinger's equation (Equation 6.2).

$$H\Psi = E\Psi \tag{6.2}$$

An expression for the energy (Equation 6.3) follows from Schrödinger's equation.

$$E = \frac{\int \Psi H \Psi \mathrm{d}\tau}{\int \Psi \Psi \mathrm{d}\tau} = \frac{\langle \Psi|H|\Psi \rangle}{\langle \Psi^2 \rangle} \tag{6.3}$$

All that needs to be done is to calculate the coefficients c_i in the expression for Ψ (Equation 6.1). This can be done by invocation of the *variation principle*.

6.4.1.1 The Variation Principle. The variation principle is that the energy calculated using an approximate wavefunction will never be lower than the true energy. Therefore, the wavefunction which gives the lowest energy will be the best.

The two integrals in Equation 6.3 can be expanded to give an expression for the energy of a system in terms of the atomic orbitals, ϕ_i, and the coefficients, c_i.

$$\int \Psi\Psi d\tau = \int (c_1\phi_1 + c_2\phi_2 + \ldots + c_n\phi_n)(c_1\phi_1 + c_2\phi_2 + \ldots + c_n\phi_n)d\tau$$

$$= \int c_1c_1\phi_1\phi_1 d\tau + \int c_1c_2\phi_1\phi_2 d\tau + \ldots + \int c_1c_n\phi_1\phi_n d\tau$$

$$+ \int c_2c_1\phi_2\phi_1 d\tau + \ldots + \int c_nc_n\phi_n\phi_n d\tau$$

$$\int \Psi H\Psi d\tau = \int (c_1\phi_1 + c_2\phi_2 + \ldots + c_n\phi_n)H(c_1\phi_1 + c_2\phi_2 + \ldots + c_n\phi_n)d\tau$$

$$= \int c_1c_1\phi_1 H\phi_1 d\tau + \int c_1c_2\phi_1 H\phi_2 d\tau + \ldots + \int c_1c_n\phi_1 H\phi_n d\tau$$

$$+ \int c_2c_1\phi_2 H\phi_1 d\tau + \ldots + \int c_nc_n\phi_n H\phi_n d\tau$$

(6.4)

Equations 6.4 look horrendous, but the difficulty lies mainly in the very large number of terms involved. Going through them a one at a time, repeating the same procedures, is something that computers do very well. The variation principle suggests that the choice of c_i which gives the lowest value of E will be the best. In order to find this, the derivatives of the expression for E are required, with respect to each c_i, because at the minimum energy value all of these derivatives will be zero. The equations can be made to look simpler by abbreviating the integrals as defined in Equation 6.5.

$$\int \phi_i H\phi_i d\tau = \alpha_i$$

$$\int \phi_i H\phi_j d\tau = \beta_{ij}$$

$$\int \phi_i \phi_j d\tau = S_{ij}$$

$$\int \phi_i \phi_i d\tau = 1$$

(6.5)

By taking the derivative of Equation 6.3 with respect to each c_i, and setting each of the resulting expressions equal to zero, the n equations summarised in alternative but equivalent forms in Equations 6.6 may be obtained. A number of lines of working and rearrangement are omitted here.

$$(\alpha_1 - E)c_1 + (\beta_{12} - E\ S_{12})c_2 + \ldots + (\beta_{1n} - E\ S_{1n})c_n = 0$$
$$(\beta_{21} - E\ S_{21})c_1 + (\alpha_2 - E)c_2 + \ldots + (\beta_{2n} - E\ S_{2n})c_n = 0$$
$$\cdots \quad \cdots \quad \cdots \quad \cdots \quad \cdots \quad \cdots \quad \cdots$$
$$(\beta_{n1} - E\ S_{n1})c_1 + (\beta_{n2} - E\ S_{n2})c_2 + \ldots + (\alpha_n - E)c_n = 0 \qquad (6.6)$$

$$\begin{pmatrix} \alpha_1 - E & \beta_{12} - E\ S_{12} & \ldots & \beta_{1n} - E\ S_{1n} \\ \beta_{21} - E\ S_{21} & \alpha_2 - E & \ldots & \beta_{2n} - E\ S_{2n} \\ \cdots & \cdots & \cdots & \cdots \\ \beta_{n1} - E\ S_{n1} & \beta_{n2} - E\ S_{n2} & \ldots & \alpha_n - E \end{pmatrix} \begin{pmatrix} c_1 \\ c_2 \\ \cdots \\ c_n \end{pmatrix} = \begin{pmatrix} 0 \\ 0 \\ \cdots \\ 0 \end{pmatrix}$$

There are n of these equations, which are called secular equations, and n coefficients c_i to find. However, it is clear that one solution to these equations is for all the coefficients c_i to be zero. This is not a useful solution. Systems of equations of this type only have non-zero solutions if the determinant of the matrix formed by the coefficients is zero. This will only be true for particular values of E (matrix eigenvalues if $S_{ij} = 0$) of which there will be n. This is very useful, because it means that the equations will not only provide values for the coefficients of the wavefunction, c_i, but also a series of possible values for the energy, E. Finding the eigenvalues of the matrix will give a series of n values for E for which the equations can be solved, and these correspond to the energies of the orbitals. Each value of E will correspond to a set of values for the coefficients c_i (the eigenvectors of the matrix) and these will describe the different molecular orbitals, expressed as linear combinations of the bases ϕ_i. It can be shown that these eigenvectors will be orthogonal to each other. All that needs to be done is to work out all of the integrals necessary to find all of the values for α_i, β_{ij} and S_{ij}. This requires about n^2 integrals to be calculated.

The values of S_{ij} can be found since expressions for the atomic wavefunctions ϕ_i are available, and are qualitatively illustrated in Figure 6.1. The integrals containing a Hamiltonian operator, $\int \phi_i\ H\ \phi_i\ \mathrm{d}\tau$, require an expression for the Hamiltonian, which obtains a value for the energy from a wavefunction, by adding up the contributions from the kinetic energy of the corresponding particle, and all of the potential energy terms, which arise from the electrostatic repulsion between each electron and all the other electrons and the nuclei. This is difficult, because it suggests it is only possible to solve the equation for one electron if the positions of all the other electrons are already known. This rather fundamental problem can be solved by using the Born–Oppenheimer approximation and the Hartree–Fock method, which are described below. However, it is possible to find useful information about molecules without evaluating the integral $\int \phi_i\ H\ \phi_i\ \mathrm{d}\tau$ and one approach to doing this is Hückel theory.

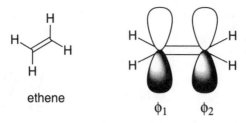

ethene

ϕ_1 ϕ_2

Figure 6.4 *Ethene π-orbital*

6.4.2 Hückel Theory

These equations can be made much more straightforward by using extreme approximations. One such approach is called Hückel Theory (Hückel, 1931), and this only takes account of π-electrons in delocalised systems. For example, consider ethene, Figure 6.4. The Hückel calculation only takes account of the two p orbitals on the sp^2 carbons, which are labelled ϕ_1 and ϕ_2.

The equivalents of Equations 6.6 may now be constructed, and the result is fairly simple, as only two orbitals are involved.

$$(\alpha_1 - E)c_1 + (\beta_{12} - E\,S_{12})c_2 = 0$$

$$(\beta_{21} - E\,S_{21})c_1 + (\alpha_2 - E)c_2 = 0 \qquad (6.7)$$

The quantities α_i are the integrals of the p orbitals, both of which are the same. A single value, α, can therefore be used for both, and corresponds to the energy of an isolated p orbital. β_{12} and β_{21} are equal by symmetry, and so will be called β. S_{12}, the overlap of the two p orbitals, will be set equal to zero. The equations simplify further to:

$$(\alpha - E)\,c_1 + \beta\,c_2 = 0$$

$$\beta\,c_1 + (\alpha - E)\,c_2 = 0 \qquad (6.8a)$$

or, writing in a matrix form

$$\begin{pmatrix} \alpha - E & \beta \\ \beta & \alpha - E \end{pmatrix}\begin{pmatrix} c_1 \\ c_2 \end{pmatrix} = \begin{pmatrix} 0 \\ 0 \end{pmatrix} \qquad (6.8b)$$

These equations can only have non-zero values for c_1 and c_2 if the determinant of the left hand side is zero, *i.e.* $(\alpha - E)^2 - \beta^2 = 0$. This gives the results:

$$E = \alpha + \beta \text{ and } c_1 = c_2 \quad \text{ or } \quad E = \alpha - \beta \text{ and } c_1 = -c_2 \qquad (6.9)$$

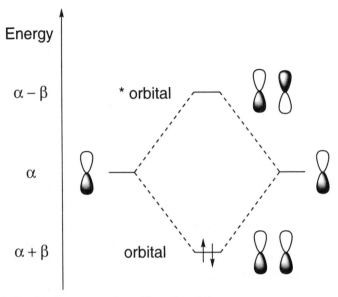

Figure 6.5 *Graphical representation of Equations 6.7*

This simple calculation has given the result that two p orbitals will interact to form two molecular orbitals, one higher in energy than the atomic orbital from which they were formed, and the other lower in energy. There are two electrons in the π-system of ethene, and these will both go in the lowest energy orbital available (Figure 6.5).

In fact, S_{ij} (Equation 6.7) is not exactly zero. This has the result that the antibonding orbital, π^*, will be raised in energy, relative to the isolated atomic orbitals, slightly more than the bonding, π, orbital is lowered in energy.

6.4.3 Born–Oppenheimer approximation

The Hückel approximation is useful for some cases, but it is often necessary to consider the evaluation of the integrals $\int \phi_i H \phi_i \, d\tau$ more rigorously, which means it is necessary to take account of how every particle affects every other particle. This is a very difficult problem, which is usually approached using approximations. The first of these is the Born–Oppenheimer approximation. This is the idea that electrons move very much faster than atomic nuclei, and so the movement of electrons can be considered independently of the movement of the nuclei. This is a very good approximation, even in the worst possible case where the nucleus in question is a proton, which is less than two thousand times the weight of an electron. It is not necessary to introduce corrections for this major simplification in most cases. Although moving any electron in a system will affect every other electron and all the atomic nuclei as well, the Born–Oppenheimer approximation means that it is possible to treat the nuclei as unmoving objects, and only consider the effects on the other electrons.

6.4.4 Hartree–Fock

The second important approximation is the Hartree–Fock self-consistent field method. The energy of a particular electron depends on the electric fields produced by the atomic nuclei and by all the other electrons. If the wavefunctions for all the electrons except one are known, then it is possible to calculate the wavefunction of the remaining electron. If none of the wavefunctions are known, then this poses a major problem. The problem can be compared with that of laying out a new flower bed. A number of plants will be arranged in the bed, but the position of each plant will depend on all of the others, because they should be related to each other as well as to the boundaries of the flower bed. This problem can be solved by arranging the plants in something like the right position, and then moving each of them, one at a time, until a good arrangement is found. The Hartree–Fock self-consistent field approach works in a similar way. First, the effects of all the electrons except one are averaged. The average is necessary because the electrons are moving around. It is now possible to find how the one remaining particle will move in the average electric field created by all the rest. This movement is then averaged, and the next electron is selected. The process is repeated until a self-consistent solution is obtained. This approach requires an initial guess for the wavefunctions, which can be found from a Hückel approach.

Since the repulsion between pairs of electrons must be calculated, the difficulty of working out these integrals $\int \phi_i H \phi_i \, d\tau$ is approximately proportional to n^2. In Section 6.4.1.1 it was shown that the number of integrals which must be calculated scales as n^2 as well, so the difficulty of obtaining a solution to these equations should scale as n^4. This is a useful rule of thumb, although not a precise measure of the difficulty of such calculations.

A further approximation is often used. If a system has a multiplicity of zero, which means that all the electrons are paired, then the electrons may be restricted to moving in pairs. This is called restricted Hartree–Fock (RHF), and is the best method for such systems. If unpaired electrons are present, unrestricted Hartree–Fock (UHF) must be used.

It is not easy to do this, because particles are not static points, but waves oscillating or charges moving in their orbitals, and these orbitals are rather difficult to describe. The usual approach to describing orbitals is to use a set of basis functions as building blocks. Simple building blocks will make the calculation rapid, but less accurate. More sophisticated building blocks will slow the calculation down, but should lead to higher accuracy. These building blocks are referred to as *basis sets*, and a wide variety have been developed.

How is it possible to decide which set of building blocks, which basis set, will be best for a particular problem? The variation principle is a very useful theorem in these circumstances. It states that the energy calculated using an approximate wavefunction will never be lower than the true energy. Basis sets are approximations to the true wavefunction, and so the energies derived from them will always be too high. Better basis sets will give lower energies.

6.4.4.1 Basis Sets. Two sorts of basis functions have been widely used. Slater-type atomic orbitals (STO), which were introduced in the 1930s (Slater, 1930), provide reasonable representations of atomic orbitals. However, they are rather difficult to manipulate mathematically. Gaussian-type atomic functions have now largely superseded Slater orbitals. A single Gaussian function does not provide a very good representation of an atomic orbital, but the functions are easy to manipulate because the product of two Gaussians is another Gaussian. Combinations of Gaussians can be used to make good approximations to atomic orbitals.

Some simple Gaussian basis sets mimic Slater-type atomic orbitals. For example. STO-3G is a basis set which uses three Gaussian functions to form each Slater-type orbital. This is a slightly cruder model than STO-4G, which uses four Gaussians to form each Slater-type orbital, and so is likely to give somewhat higher energies for any system.

Split-valence basis sets are commonly designated *a-bc*G, where each of *a*, *b*, and *c* correspond to the number of Gaussian functions used in each part of the model. The larger these numbers, the more precise the basis set, but the more time will be required to complete a calculation. 3-21G is the smallest basis set in common use. 6-31G produces better answers, if the greater computational expense of using it is acceptable.

These basis sets do not allow for the polarisation of orbitals. This can be added into the basis sets, and is usually designated by asterisks. Thus 6-31G* has polarisation functions on the non-hydrogen atoms, and 6-31G** has polarisation functions on all atoms. Calculations involving anions can require the use of additional diffuse functions, which are indicated by a + sign. Thus 6-31 + G* includes both polarisation functions and also diffuse functions. There are also other naming schemes for basis sets which may be considerably more flexible than 6-31 + G*.

The time required to perform a molecular orbital calculation increases approximately as N^4, where N is the number of basis functions. Increasing the size of the basis set will have a dramatic effect on the time required for a calculation.

6.4.4.2 Electron Correlation. The Hartree–Fock method makes calculations very much easier, but ignores the Pauli principle, which says two electrons cannot be in the same place. In other words, electrons do not move in average electric fields, but their movement is in some way correlated. The usual approach to this problem is to make corrections to the Hartree–Fock approximation to allow for this, rather than to use a completely different method. The corrections can either be based on the consideration of excited electronic states (Configuration Interaction or CI), or by a perturbation theory correction to the Hartree–Fock method. The procedure of Møller and Plesset is commonly used, and the level of perturbation theory is indicated by a number. Thus an MP2 correction should increase the accuracy of a Hartree–Fock calculation. An MP3 or an MP4 correction should give greater accuracy, but at the expense of greater computational cost.

The results of these approximations are shown in Figure 6.6. A precise solution to Schrödinger's equation could be found, in principle, by using an infinitely flexible basis set, and making a complete correction for the effects of electron correlation. Neither of these things is possible in practice. The figure has the names of a few simple basis sets, and a few corrections for electron correlation.

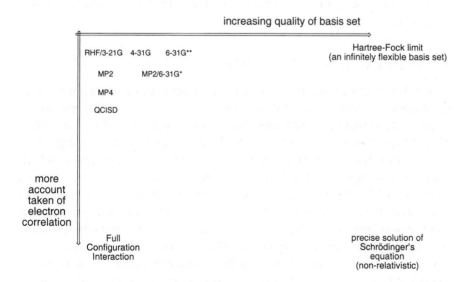

Figure 6.6 *Precision increases with basis set size and corrections for electron correlation*

Electrons may move close to the speed of light, and so relativistic effects may be important. In practice, this is most important for electrons in the cores of heavier elements, and leads to a contraction in the size of the orbitals of the inner electrons.

The Hartree–Fock method gives a description of the electronic structure of a molecule, and this produces an energy for the molecule. The energy is likely to be very large and negative (although the zero of energy may vary between different molecular orbital programs), as it may relate to the difference in energy between the molecule and isolated particles. The energy is often expressed in Hartrees, which are a convenient unit for this sort of calculation. One Hartree corresponds to $2626 \, kJ \, mol^{-1}$.

Such a calculation will give the energy of an arrangement of atoms, just as a force field will, the only difference being that the calculation is very much more complex. It is possible, therefore, to use the energy calculated by a molecular orbital calculation to minimise molecules, to perform conformation searching and to run molecular dynamics simulations.

Molecular mechanics models require a lot of information about the bonds in molecules, the strength of all the bonds, the difficulty of bending the bonds, and so on. Molecular orbital calculations can find an energy for a system simply from the atom positions, and the basis set for the individual atoms. There is no need to specify which atoms are bonded to which (there are molecules for which this is very difficult to define!). Molecular orbital calculations can, therefore, be applied to systems for which very little is known, unlike molecular mechanics which needs a force field.

6.5 SEMI-EMPIRICAL METHODS

When the ideas of *ab initio* molecular orbital calculations were first developed, it was clear that the number of arithmetical operations required to investigate even the simplest of systems would be very large. Drastic approximations were required. One approach was to neglect all electrons except the valence electrons, to decide that some orbitals which only just overlapped did not overlap, and to calculate the interactions of other orbitals through parameters. This is called a semi-empirical approach. Parameters are required for every element in a system, but not for bond lengths, bond angles, *etc.* The first semi-empirical approach was CNDO (Complete Neglect of Differential Overlap: Pople *et al.*, 1965; Pople and Segal, 1965), and the parameters required were developed from *ab initio* calculations. Considering the great simplifications that were used, the results were rather good, and could be calculated quickly. A similar approach was tried by Dewar (Bingham, *et al.*, 1975) and called MINDO/3. Dewar followed the alternative philosophy of parameterising the method from experimental data rather than from *ab initio* calculations. This program was developed and called MNDO (Dewar and Thiel, 1977) and then AM1 (Dewar *et al.*, 1985). PM3 is an alternative development (Stewart, 1989). These semi-empirical methods are available in a package called MOPAC (Stewart, 1990), as well as in a number of other commercial and academic programs.

As computers became more powerful, and the calculation of all the necessary integrals began to become possible for more than the very smallest of molecules, Pople abandoned the semi-empirical approach in favour of slower but more accurate *ab initio* methods (Pople, 1975). Dewar continued the semi-empirical approach, and these programs are very widely used for a wide variety of studies. Dewar's final publication on the subject suggested that a hybrid semi-empirical *ab initio* approach may be best (Dewar *et al.*, 1993).

The chief advantage of semi-empirical molecular orbital programs over *ab initio* molecular orbital programs is their speed. The simplifications to the integrals have the result that the time required for a calculation increases only as N^3, and so quite large molecules can be studied directly. There is also a subsidiary advantage. Because the methods are parameterised against experimental results, and the experimental results include the effects of electron correlation, some allowance for this effect is implicit in the calculations. This can also be seen as a disadvantage, because it is not clear how great an allowance is being made for this effect, and so it is very difficult to assess the errors in the method.

6.6 DENSITY FUNCTIONAL THEORY

A new approach to calculating the electronic structure of molecules is being developed, which is based on calculation of the electron density rather than of molecular orbitals (Parr and Yang, 1995). This technique may mean that it will become possible to calculate the electronic properties of much larger systems than is presently possible using molecular orbital theory, because it may scale linearly in n rather than as n^4 which is approximately followed by molecular orbital methods. For small systems, however, molecular orbital methods are faster.

6.7 USING MOLECULAR ORBITAL THEORY

This chapter may have seemed rather heavy going, so far. Fortunately, however, molecular orbital theory can be quite easy to use. Just as for a molecular mechanics calculation, a molecular orbital calculation must be given the geometry of a molecule for it to begin. Unlike molecular mechanics, there is no need to specify bonds between atoms, as the molecular orbital calculation does not take account of bonds directly. Indeed, it can be rather hard to define exactly where chemical bonds are, when the computer is dealing with clouds of electron density! A molecular orbital calculation must also be given the charge on the system, and its multiplicity, which is a measure of the number of unpaired electrons in the system. Most organic molecules are singlets, which means they have no unpaired electrons.

Once a molecule has been created for a molecular orbital calculation (this may often be done by using molecular mechanics to obtain an approximate idea of the structure), and a basis set has been chosen, a molecular orbital program will perform all the calculations outlined in the preceding sections, and produce an energy, or a minimised structure, as required. The procedure is not so different from a molecular mechanics calculation, except that the amount of computer power required is much greater. The amount of information which may be obtained from a molecular orbital calculation is also greater, because the arrangement of the electrons in the molecule has now been considered.

6.8 USEFUL INFORMATION FROM MOLECULAR ORBITAL THEORY

6.8.1 Energy

An important result from a molecular orbital calculation is the absolute energy of the system. The energies tend to be large and negative, and so differences in energies between different conformations of one system or between different systems are usually the most useful quantities to calculate. Different basis sets, or different semi-empirical methods, will give very different absolute values for energy, so it is important always to compare like with like.

Because an energy has been found for a particular arrangement of atoms, it is also possible to minimise the geometry, perform conformation searches or do molecular dynamics simulations, just as described in earlier chapters. These calculations will, however, be very much slower than similar calculations using molecular mechanics. Unlike molecular mechanics, molecular orbital theory can deal with bonds forming and breaking, and so it can be used to study reactions as well as ground states. This makes it an enormously powerful tool. When a structure is minimised using molecular mechanics, the result is usually a minimum. Exceptions to this do occur, and one was discussed in Q3.2. Molecular orbital studies may be much more interested in saddle points, because they represent crucial steps in a reaction pathway, rather than just a peculiar high energy structure, as is usually the case for molecular mechanics studies. Testing that a stationary point is truly a minimum, rather than a transition structure or even a maximum, has a correspondingly higher importance in molecular orbital calculations. The test is carried out using the Hessian matrix, which comprises the second derivatives for the energy with respect to the coordinates of the atoms.

6.8.2 Molecular Orbital Energy Levels

The description of molecular orbitals which comes from a molecular orbital calculation may provide useful information about a molecule's reactivity. In particular, details of the energy and the location of the HOMO (Highest Occupied Molecular Orbital) can be found, and so it is possible to estimate how nucleophilic a molecule will be, and which parts of the molecule are most likely to attack an electrophile. Koopmans' theorem (Koopmans, 1934) states that it is reasonable to assign the HOMO energy as the ionisation potential of the system, because removing an electron from a closed shell system will not cause the orbitals to rearrange very much.

Significantly, there is no parallel result connecting the LUMO (Lowest Unoccupied Molecular Orbital) with the electron affinity. Molecular orbital calculations find the energies of molecular orbitals by calculating the interactions between electrons. Vacant orbitals have no electrons, and so receive less attention. Molecular orbital calculations produce energies and distributions for the LUMO of a system as well as the HOMO, but the result is much less reliable.

6.8.3 Charge Distribution

One of the difficulties with parameterising a molecular mechanics force field is finding values for the partial charge on each atom. Because molecular orbital theory calculates the electron distribution of a system, it can be used to find values for these partial charges. However, electron density is spread out over the whole molecule, and it is not obvious how to decide which parts of the electron density should be assigned to which atoms. Mulliken (Mulliken, 1955) and Löwdin (Löwdin, 1955) have both developed expressions which can be used for this purpose. Approximating a continuous electron distribution by atom-centred point charges is a very severe approximation, and it is not clear that a reasonable protocol for dividing the electron density between atoms will necessarily give the best results.

It is possible to divide up electron density by consideration of zero flux surfaces (Bader, 1985). One way to imagine this is to consider introducing a tiny classical negative charge into a molecular structure, and considering how it would behave, assuming it is too small to affect the molecule or its electron distribution. The test charge would always fall towards a nucleus, unless it was precisely positioned between two or more nuclei, such that the attraction to each was equal. Such positions define the zero flux surfaces, and these divide up electron density in a satisfying way. This approach can also be used to decide whether or not two atoms are bonded to each other.

An alternative approach to divide up electron density has also been developed (Singh and Kollman, 1984; Chirlian and Francl, 1987; Breneman and Wiberg, 1990). Atom-centred point charges generate an electrostatic field around a molecule, which approximates the electrostatic field produced by the electrons in their orbitals. Molecular orbital theory can be used to calculate values for this field at points around the molecule, and then atom-centred point charges can be chosen to reproduce it as well as possible. The approach, known as Electrostatic Potential or ESP derived charges, appears to give good results for the parameterisation of force fields. Semi-empirical methods appear to give similar results to *ab initio* methods (Ferenczy *et al.*, 1990).

6.9 KEY POINTS

- *Ab initio* molecular orbital methods (GAUSSIAN, CADPAC, GAMESS, Spartan, *etc.*) provide a way of calculating molecular properties based only on a knowledge of the atoms.
- *Ab initio* methods can only be used for fairly small systems, because they rapidly become more expensive, both in terms of time and computer power, as molecules get larger.
- In general, the longer the name, the better the method:
  QCISD(T)/6-311++G(3df,3pd)//QCISD(T)/6-311++G(3df,3pd)
  is excellent, but expensive;
  MP2/6-31G**//RHF/6-31G* is generally acceptable for structure calculation;
  RHF/3-21G//3-21G is commonly used.
- Semi-empirical molecular orbital methods (MOPAC) are faster than *ab initio* but less accurate. AM1 and PM3 are the most sophisticated models in general use, superseding MNDO, MINDO/3, CNDO and numerous others.

QUESTIONS

Q6.1 Hückel calculations

Do a Hückel calculation on the allyl cation ($^+CH_2CHCH_3$) and the allyl anion ($^-CH_2CHCH_3$). What is the electron distribution? Compare the energies of the orbitals with the corresponding calculation for the cyclopropenyl cation and anion. Does the Hückel calculation give any idea of aromatic stabilisation?

Do semi-empirical and *ab initio* calculations on the same systems. Are these consistent with the Hückel calculations and with each other?

Q6.2 Redox potentials

The oxidation potentials (removal of an electron) of benzene, fulvene, naphthalene, azulene, anthracene and perylene (Figure 6.7) have been measured. Is it possible to calculate these values using molecular orbital theory? How well do the calculations correlate with the experiments? What about reduction potentials?

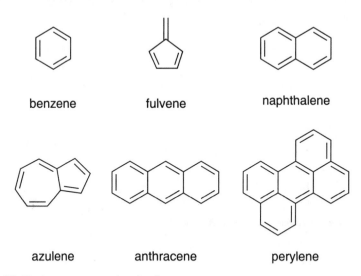

benzene   fulvene   naphthalene

azulene   anthracene   perylene

Figure 6.7 *Various unsaturated molecules*

Q6.3 Diels–Alder reactions

The Diels–Alder reaction of methoxybutadiene with acrolein (Figure 6.8) preferentially gives the *ortho* adduct. Can molecular orbital calcula-

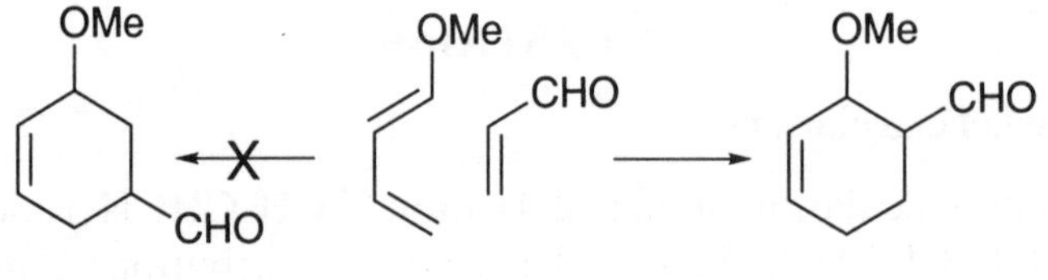

Figure 6.8 *A Diels–Alder reaction*

tions can be used to rationalise this result? Is it also possible to calculate which diastereoisomer should be formed?

Q6.4 Charge distribution

How is charge distributed over a carbonyl group? What about an ester? An amide?

Q6.5 Reaction selectivity

Which of the isomers shown in Figure 6.9 is thermodynamically preferred? Does molecular mechanics also give the correct result? Can you account for the differences?

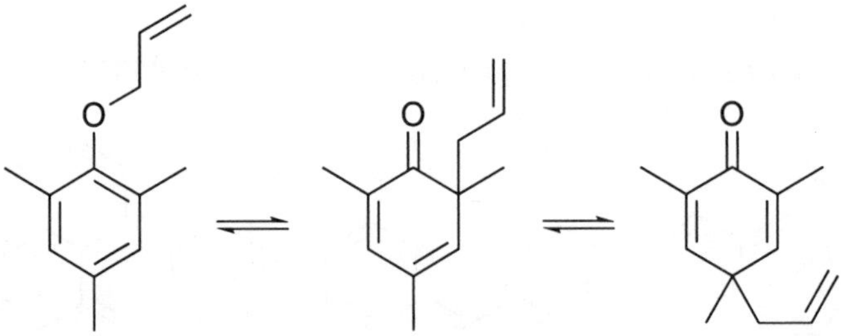

Figure 6.9 *Rearrangement of an allyl phenol*

Q6.6 A cation rearrangement

Acetolysis of tosylated nopol could give any of a number of products (Figure 6.10). What do you think is most likely to form? What would be the best calculation to do to investigate this system?

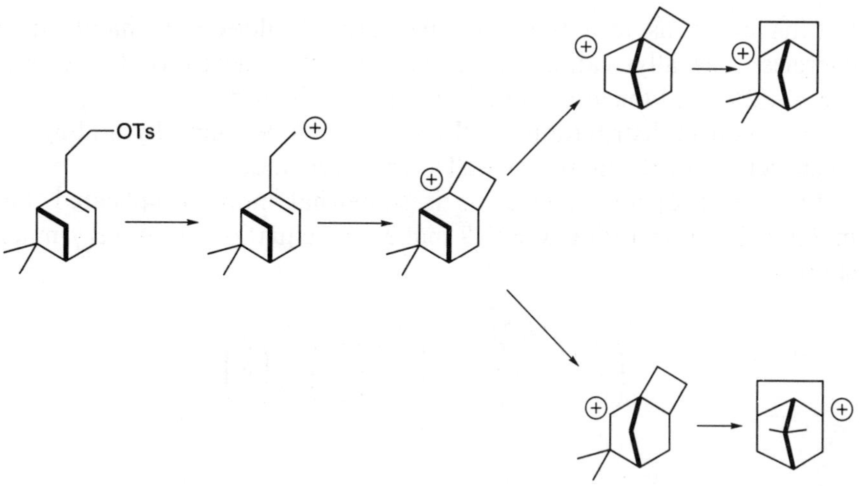

Figure 6.10 *Cationic rearrangements*

DISCUSSION

D6.1 Hückel calculations

A Hückel calculation on an allyl system:

$$\begin{pmatrix} \alpha - E & \beta & 0 \\ \beta & \alpha - E & \beta \\ 0 & \beta & \alpha - E \end{pmatrix} \begin{pmatrix} c_1 \\ c_2 \\ c_3 \end{pmatrix} = \begin{pmatrix} 0 \\ 0 \\ 0 \end{pmatrix}$$

Substituting $x = \dfrac{\alpha - E}{\beta}$ gives

$$\begin{pmatrix} x & 1 & 0 \\ 1 & x & 1 \\ 0 & 1 & x \end{pmatrix} \begin{pmatrix} c_1 \\ c_2 \\ c_3 \end{pmatrix} = \begin{pmatrix} 0 \\ 0 \\ 0 \end{pmatrix}$$

This reduces to

$$x^3 - 2x = 0 = x(x^2 - 2)$$

whence

$$x = 0; x = \pm\sqrt{2}$$

Thus for an allyl system

$$E = \alpha + \sqrt{2}\beta; \alpha; \alpha - \sqrt{2}\,\beta$$

The approximations in the Hückel calculation have the result that putting electrons into the molecular orbitals does not change their energies. The allyl cation, therefore, with two electrons, has energy $2(\alpha + \sqrt{2}\beta)$, and the allyl anion's energy is 2α higher.

The electron distribution for the system can be found by finding the eigenvectors for the matrix as well as the eigenvalues.

The cyclopropenyl calculation is somewhat more complicated, but making the substitution $x = \frac{\alpha-E}{\beta}$ makes the equations look very much simpler.

$$\begin{pmatrix} \alpha - E & \beta & \beta \\ \beta & \alpha - E & \beta \\ \beta & \beta & \alpha - E \end{pmatrix} \begin{pmatrix} c_1 \\ c_2 \\ c_3 \end{pmatrix} = \begin{pmatrix} 0 \\ 0 \\ 0 \end{pmatrix}$$

Substituting $x = \dfrac{\alpha - E}{\beta}$ gives

$$\begin{pmatrix} x & 1 & 1 \\ 1 & x & 1 \\ 1 & 1 & x \end{pmatrix} \begin{pmatrix} c_1 \\ c_2 \\ c_3 \end{pmatrix} = \begin{pmatrix} 0 \\ 0 \\ 0 \end{pmatrix}$$

whence

$$x^3 - 3x + 2 = 0$$

Note that $x = 1$ is a solution to this equation, so:

$$(x - 1)(x - 1)(x + 2) = 0$$

Thus for a cyclopropenyl system

$$E = \alpha + 2\beta; \alpha - \beta; \alpha - \beta$$

The energy of the cation is now $2(\alpha + 2\beta)$ and the anion is $2(\alpha - \beta)$ higher. This means that the π-electrons are lower in energy for cyclopropenyl than for allyl, if the system is a cation, and higher in energy if it is an anion. The cyclic cation has two electrons, and so should show aromatic stabilisation. The anion should be anti-aromatic and so destabilised. This is consistent with the Hückel calculation. The same calculation can be repeated for butadiene. In this case, neutral cyclobutadiene should be antiaromatic and the double charged system aromatic. (The calculation for benzene is similar, but rather time consuming!)

For butadiene

$$
\begin{pmatrix} x & 1 & 0 & 0 \\ 1 & x & 1 & 0 \\ 0 & 1 & x & 1 \\ 0 & 0 & 1 & x \end{pmatrix} \begin{pmatrix} c_1 \\ c_2 \\ c_3 \\ c_4 \end{pmatrix} = \begin{pmatrix} 0 \\ 0 \\ 0 \\ 0 \end{pmatrix}
$$

$$x^4 - 3x^2 + 1 = 0$$

substituting $y = x^2$

$$y^2 - 3y + 1 = 0$$

$$y = \frac{3 \pm \sqrt{5}}{2}$$

$$x = \pm\sqrt{\frac{3 \pm \sqrt{5}}{2}} = \pm\frac{-1 \pm \sqrt{5}}{2}$$

$$E = \alpha \pm 1.618\,\beta;\, \alpha \pm 0.618\,\beta$$

and for cyclobutadiene

$$
\begin{pmatrix} x & 1 & 0 & 1 \\ 1 & x & 1 & 0 \\ 0 & 1 & x & 1 \\ 1 & 0 & 1 & x \end{pmatrix} \begin{pmatrix} c_1 \\ c_2 \\ c_3 \\ c_4 \end{pmatrix} = \begin{pmatrix} 0 \\ 0 \\ 0 \\ 0 \end{pmatrix}
$$

$$x^4 - 4x^2 = 0$$

$$x^2(x^2 - 4) = 0$$

$$E = \alpha;\, \alpha;\, \alpha - 2\beta;\, \alpha + 2\beta$$

D6.2 Redox potentials

Oxidation and reduction potentials are conventionally measured in electron volts. Conversion factors are given in Appendix A.1. The AM1 Hamiltonian and MOPAC were used to minimise the structures and to calculate the orbital energies.

Oxidation potentials are the energetic cost of removing an electron from the molecule's HOMO. These correlate well with the HOMO energy (which is negative because it is measured in the opposite direction), in accord with Koopmans' theorem (Table 6.1). There is no complementary rule relating LUMO energies to reduction potentials, even though these correspond to putting an electron into the LUMO.

Table 6.1 *Oxidation and reduction potentials (eV)*

Molecule	Oxidation potential	HOMO energy	Reduction potential	LUMO energy
benzene	9.24	−9.65	–	0.55
fulvene	8.36	−9.06	–	−0.44
naphthalene	8.14	−8.71	–	−0.26
azulene	7.38	−8.02	0.69	−0.87
anthracene	7.44	−8.12	0.57	−0.84
perylene	6.96	−7.86	0.97	−1.16

Experimental data from *CRC Handbook of Chemistry and Physics*

LUMO energies are a poor guide to reduction potentials and electron affinities.

D6.3 Diels–Alder reactions

The MOPAC implementation of AM1 was used to calculate the HOMO and the LUMO energies and coefficients. The methoxybutadiene component has a HOMO of −8.77 eV and a LUMO of 0.56 eV. (A full conformation search was not performed for this molecule. The cautious chemist would certainly have done this, to investigate how the molecular orbitals alter with rotation of the methoxy group. The *S-cis* conformation must be present, however, for the reaction to occur at all.) Acrolein has a HOMO of −10.70 eV and a LUMO of −0.14 eV. The closest matched HOMO/LUMO pair is between the HOMO of the diene and the LUMO of the dienophile, and the orbitals which are closest in energy will interact most strongly. These are the orbitals which need to be considered.

Is it reasonable to trust the LUMO energy in this case? The LUMO energy will certainly be calculated with a lower precision than the HOMO energy. The LUMOs in this example are rather similar in energy and the HOMOs differ by a greater amount. It is likely that the HOMO with the highest energy will be the important one, and this leads to the same conclusion, even if the LUMO energies are very inaccurate. The molecular orbital calculation also finds the coefficients of the orbitals on each atom, and so it is possible to check that the HOMO and the LUMO both correspond to the π-orbitals of the system. In this example, all four orbitals correspond to the π-orbitals, as expected, and the coefficients are given in Figure 6.11.

The atoms with the largest coefficients will interact most strongly, so it is clear that the lowest carbon atom of the butadiene (as drawn in Figure 6.11) will react with the lowest carbon atom of the acrolein. This

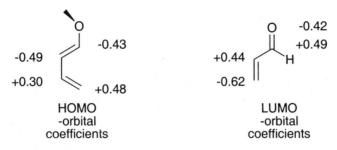

HOMO
-orbital
coefficients

LUMO
-orbital
coefficients

Figure 6.11 *Orbital coefficients for the Diels–Alder reaction*

represents a negative coefficient interacting with a positive coefficient, so all favourable interactions must have the same combination of signs. This means that the other end of the C=C bond of acrolein will interact favourably with the other end of the butadiene system, to do a Diels–Alder reaction. This will lead to the observed regiochemistry for the products.

The stereochemistry of the product will depend on whether the reaction goes through an *exo* or an *endo* transition state. The carbonyl carbon component of the orbital will interact favourably with one of the central atoms of the butadiene system, and so the *endo* product will be favoured, again, consistent with experiment.

D6.4 Charge distribution

ESP charges are often smaller than Mulliken charges (Figure 6.12), particularly for atoms buried in the middle of a molecule.

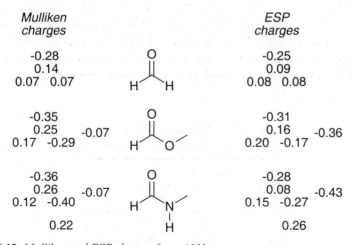

Figure 6.12 *Mulliken and ESP charges from AM1*

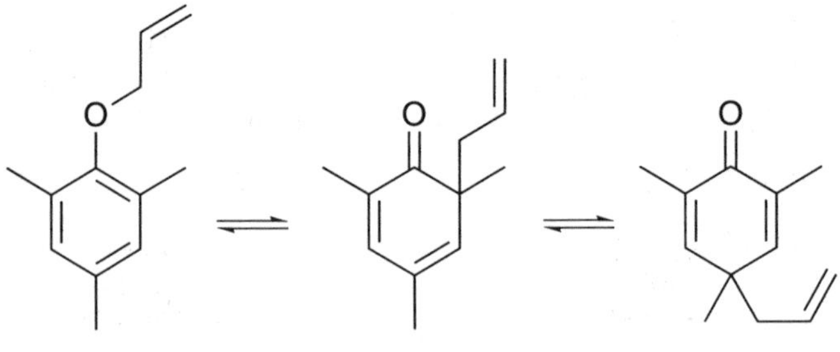

Figure 6.13 *Rearrangement of an allyl phenol*

D6.5 Reaction selectivity

A molecular mechanics conformation search was performed on the three isomers shown in Figure 6.13. All of the conformations from each search were then reminimised with AM1. The energies of the global minima are given in Table 6.2.

Molecular mechanics obtains a dramatically different result to semi-empirical molecular orbital theory. Molecular mechanics suggests that **A** should be barely present, and semi-empirical methods suggest that it should be the major product. Molecular mechanics is likely to be unreliable in this sort of case, because the isomers being compared are not sufficiently closely related. In particular, the conjugated systems are rather different, and only **A** is aromatic. The isomers are too dissimilar to be reliably compared using molecular mechanics.

The semi-empirical calculations, based on the molecular mechanics conformation searches, show that isomer **A** is much the lowest in energy, and so should be thermodynamically preferred. This is consistent with the experimental result (Kalberer *et al.*, 1956). Reaction conditions have been reported for the allylation of 2,4,6-trimethylphenol which give **C** exclusively. The explanation for this can be found by examining the HOMO of the 2,4,6-trimethylphenolate anion, which has its largest

Table 6.2 *MM2* and AM1 energies for global minima* $(kJ\,mol^{-1})$

Method	Isomer		
	A	B	C
MM2*	90.2	18.7	21.3
AM1	−65.7	−38.2	−32.6

coefficient on the *para*-carbon of the phenolate ring, suggesting that electrophilic addition should be kinetically favoured at this site.

D6.6 A cation rearrangement

The energies of all the intermediate cations (Figure 6.14) can be calculated using AM1, suggesting that the lower path has a lower energy barrier than the upper path, even though it requires a methylene group to migrate rather than a *gem*-dimethyl. It also leads to the lowest energy final product. This is consistent with the experimental result.

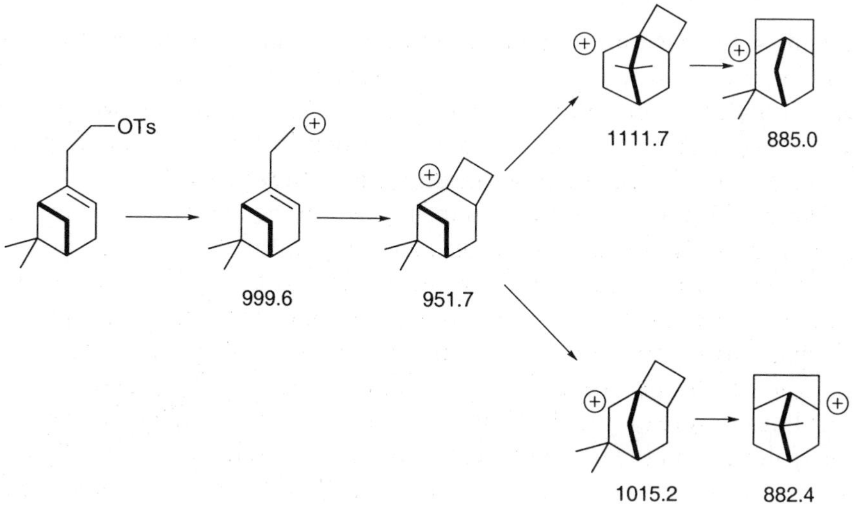

Figure 6.14 *AM1 energies of cations* (kJ mol^{-1})

CHAPTER 7

Databases

7.1 INTRODUCTION

Chemists are generating a huge amount of data – too much for any individual to analyse. A computer is a very useful tool for sorting and searching large collections of data. The same computers which are needed to perform the calculations described in the previous chapters can also be used for managing the data which they generate.

This problem has already been mentioned in the context of molecular dynamics. A dynamics simulation generates so much data that only a computer can analyse it, and produce some sort of summary for the scientist to examine. The same applies to conformation searching, for which algorithms are available to group the results of calculations, or to sort the results by energy or some other criterion. Molecular orbital calculations were originally so hard that all of them would be archived. It was easier to find the result of a calculation in an archive than to rerun it. Now, so many calculations have been done that it may be quicker to re-run a simple calculation, rather than to look up the answer.

The published literature is growing very rapidly. The number of papers published is increasing, but this is only a part of the phenomenon, because it is possible to put more data into each paper. Three-dimensional structures used not to be a part of the published literature, but now they play an important role. For example, the Cambridge Crystallographic Database (Allen and Kennard, 1993) maintains information on over 175000 structures, and the Brookhaven Protein Databank (Bernstein *et al.*, 1977) contains data on over 7000 biological macromolecules. Some journals now offer enhanced articles containing three-dimensional on-line information as well as the traditional, two-dimensional, printed page. The Royal Society of Chemistry started producing articles enhanced in this way in 1996. There is no end in sight to the increase in both the breadth of data, as assessed by the number of papers, and its depth, as new sorts of information can become available.

7.1.1 Journals and Published Data

Chemical Abstracts, which contains a short abstract of every chemistry paper that is published, is a very important resource for chemists. The size of *Chemical Abstracts*, as assessed by the number of centimetres of shelf space it occupies, is plotted against the date in Figure 7.1. The graph also shows the amount of shelf space required for the five year indices (until 1956 these were ten-year indices, so the earlier data has been split equally into two five-year sections). Both of these are increasing exponentially.

The five-year index became larger than the data it indexes before the end of the twentieth century. Extrapolating into the future (always a dangerous thing to take too literally), the five-year index will require 100 m of shelf space in about 2030. Extrapolating further (much less reliably, of course), by the year 2500 the index to the abstracts of the world's chemical literature will be growing at the speed of light. No library will be able to afford this sort of space, if the extrapolation for the

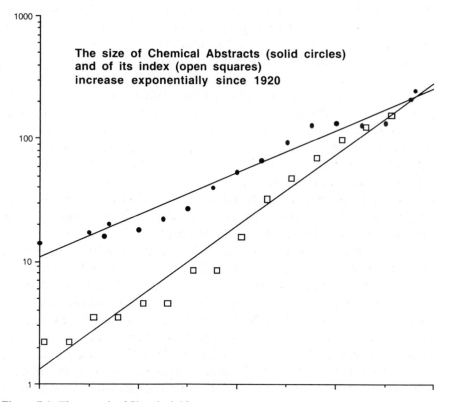

Figure 7.1 *The growth of* Chemical Abstracts

next fifty years holds true. Some other way must be found of managing this quantity of data.

This is well established within specialist subjects such as the management of crystallographic data. No printed version of the Cambridge Crystallographic Database is available, because it would not be very useful. Only a computer can manipulate such a huge quantity of data in a useful way. It is clear that the combined use of databases and molecular modelling could be very powerful, but the possibilities are only beginning to be explored.

7.2 TEXT SEARCHING

The simplest sort of database searching is based on words, and this is conceptually the same as the indexes and concordances which have been available for hundreds of years. The task is not a simple one, as there are many subtleties in how searches must be framed. For example, a database of the titles of papers, authors and references is not easy to set up.

Listing names may appear trivial at first, but there are many difficulties that arise. What should be done with double-barrelled surnames? Surnames which have several parts but are not hyphenated (*e.g.* A B Smith III; J Rebek, Jr; F de Jong)? How about names that change (*e.g.* S Gibson née Thomas)? Initials that can be abbreviated in several ways (*e.g.* P von R Schleyer)? These problems can be tackled by insisting that data are entered in a precisely defined format, separating initials, surnames and unusual features on entry, although this makes data entry much more time consuming and increases the number of errors. Anyone knowing the rules can then search a database precisely. If the rules are not sufficiently familiar (and they are likely to be complicated) then it becomes hard to frame questions. Alternatively, a sensible choice can be made at data for each difficult name. Unfortunately, different people will make different sensible choices, so different versions of the same name may well exist in a single database.

Journal titles are an easier subject, because it is possible to have a list of all the permitted journals, and to check that all new data correspond to a genuine journal. This process can be extended to check that the volume number corresponds to the year, and to check that the number of pages of the article is sensible (no articles have a negative number of pages, and few have more than a hundred). Even in this simpler case, there are decisions to be made. What to do with journals which are known by different titles in different languages, or publish different editions in different languages? What about journals which change their names?

A difficult part of the construction of a database of chemical papers is the choice of searchable keywords. Finding a consistent, globally comprehensible and orthogonal set of keywords may be a hopeless task. A less than optimal set of keywords will be useful, and give a glimpse of what might be.

7.2.1 Mark-up

It would be useful to be able to label surnames as surnames, molecule names as molecule names, and so on. Thus, if the word 'Brown' appears in a document, it would be useful to know if this were a University, a person or a residue. This can be achieved by agreeing to reserve a few special characters. A widely used choice for such special characters are the greater than and less than signs, or angle brackets. If everything inside angle brackets is a label, then the different components of a document can be marked up so that they can be distinguished. A list of the labels that should be used is also needed, because confusion could still arise as to whether it is better to use < surname > or < family name > or some other choice. SGML (Standard Generalised Mark-up Language) works in this way. The list of the labels is referred to as a 'document type definition' or *dtd*. Even if the content of a database is hard to understand, the structure of the document should be comprehensible through the mark-up.

7.2.2 Substructure Searching

A database of this type would be relatively easy to manage, because it consists entirely of words. On the whole, chemists are much more interested in structures. It is possible to work out the systematic name for a complex structure, but it is difficult and time consuming. It may also be unhelpful if the real problem is finding all molecules which contain a particular combination of functional groups. It is becoming unnecessary with the advent of substructure searching.

It has become routine to sketch a molecule, or a fragment of a molecule, on a computer screen and search a large database for anything matching this structure. This may seem to be much more complicated than searching for text, but methods of drawing chemical structures are probably more tightly defined than naming conventions, and so it is much easier to draw unambiguous structures than to search for complicated names. There may still be problems with delocalised structures, resonance forms, molecules which exist as mixtures, molecules with uncertain bonding and macromolecules, but substructure searching has made database searching much more accessible.

7.2.3 Three-dimensional Searching

The next step in complexity is three-dimensional searching. This is conceptually straightforward, if three-dimensional data are available. A search for a molecule with, for example, a carboxylic acid group between 4 Å and 5 Å from an aromatic ring is possible, but will be time consuming as a calculation will have to be done on every molecule which contains both functional groups to find if they fit the search criterion.

A more difficult problem is a three-dimensional search on a database containing two-dimensional data. This can be done using a rule based method of converting a two-dimensional description of a molecule into three dimensions (Chapter 4). Such a method can be very rapid, as it must be if a large database is to be searched. This type of method will usually only produce a single three-dimensional structure from each two-dimensional diagram. This is no problem for small and rigid molecules, but it may not be a good description of a flexible molecule. A three-dimensional description of a flexible molecule is possible, by storing the global minimum conformation and all low energy local minima. This would greatly increase the size of the database.

A three-dimensional database may be asked to produce molecules which will fit in a particular active site. The bound conformation of a molecule may not be a minimum on its potential energy surface, if the bound conformation is at an accessible energy. Simply storing a list of minima may not, therefore, be the ultimate form of three-dimensional database. A sophisticated conformation analysis program, using the methods described in earlier chapters, may need to be linked to the database to obtain reliable results.

7.3 THE WORLD WIDE WEB

The World Wide Web (WWW) is a convenient and easy way to access the internet. Many programs, called web browsers, are available which display pages of text and pictures, some of which may be highlighted. Clicking on any highlighted word, phrase or picture will download a new page connected to the word that was clicked, and this may come from a World Wide Web site anywhere in the world. Every page has a unique *Uniform Resource Locator*, or URL. The chemistry department at Cambridge's home page has a URL: http://www.ch.cam.ac.uk/ This URL provides links to thousands of chemistry sites around the world.

World Wide Web documents are written in HTML, which is a simple SGML. Despite, or perhaps because of, its simplicity, HTML has proved extremely popular. It was designed to enhance the semantics of a document, separating out headings from text, and so on. However, it can be used as a page-presentation language, and this has provided some of its appeal. A simple HTML mark up converts a dull text document into something more sophisticated, which can displayed on a wide variety of computers all around the world. There are other methods of describing page-layouts much more precisely, but with greater complexity.

The World Wide Web has had a particular impact on chemistry, because it provides a medium for the transmission of three-dimensional information, which is of crucial importance to chemists (Murray-Rust *et al.*, 1997).

7.3.1 How Popular is the World Wide Web?

The Cambridge Chemistry WWW server has run since April 1994. It now distributes about thirty thousand documents every week. A plot of the number of accesses to the server is shown in Figure 7.2. The annual dips in the line correspond to Christmas and New Year.

A straight line fits the data reasonably well, with a correlation coefficient of 0.9. This allows an estimate to be made of future use of the server. If demand for its information continues to grow linearly, it may be expected to be delivering around forty thousand documents a week in the year 2000. This sounds a large number, but it is easily within the capabilities of the workstation that has been used as a server since 1994, and was bought in 1991. If the growth is closer to exponential, the

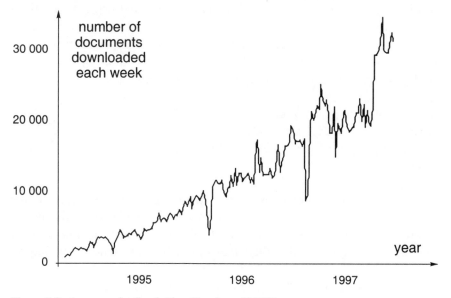

Figure 7.2 *Access to the Cambridge Chemistry WWW server*

server may have to deal with the distribution of about two hundred thousand documents a week.

The demand for information has been paralleled by the increasing complexity of the information which is available from the web server. In the summer of 1994, there were only a handful of chemistry web-servers around the world. In 1997, the number was well into four figures.

7.3.2 Simple HTML

World Wide Web pages are written in a format called HTML (Hyper-Text Markup Language), which is an example of an SGML. New versions of the *dtd* for HTML are being developed, containing new and more sophisticated instructions. This section is a brief description of the simplest instructions, which are sufficient to create useful WWW pages. Instructions are written within angle brackets, < and >, and usually come in pairs. There are not many different instructions which can be used. A well chosen, but limited, list has been a key feature of HTML. For example <H1> means Heading of First Importance. All text following such an instruction will be large, until the instruction </H1> is reached, which means that the end of the heading has been found. Most instructions come in pairs, except the <P> and <BR> commands which mean new paragraph and line break, respectively. The instructions are often called *tags*.

An important part of hypertext are *hyperlinks*. Some words will be highlighted in a page of hypertext, and clicking on them downloads another page of related information. This is achieved using instructions in angle brackets. The highlighted text is regarded as an anchor, so the instruction is *A*. The instruction also needs to tell the computer where to go if someone clicks on the highlighted text. This is done by including a URL, as a Hypertext REFerence (HREF). For example:

```
<A HREF="http://www.ch.cam.ac.uk/">
This text is all highlighted.
Click here to go to the Chemistry Department page.
</A>
```

Images can be included using the IMG tag. However, not all web browsers can display images, so it is polite to provide some alternative text which a non-image browser can display in place of the image. If the source of the image is a file called *image.gif* then the instruction would be:

```
<IMG SRC="image.gif" ALT="image of no significance
beyond decoration">
```

Table 7.1 *Hypertext with comments*

```
<HTML>                          <!--this is an HTML document!-->
<HEAD>                          <!--with a header-->
<TITLE>Title</TITLE>            <!--title line-->
</HEAD>                         <!--end of header-->
<BODY>                          <!--start of the main body of text-->
<IMG WIDTH=100 HEIGHT=50
SRC="dna.gif" ALT="DNA">        <!--an image, source file: dna.gif-->
<H1>Heading</H1>                <!--HEADING-->
<H2>Sub-heading</H2>            <!--only a sub-heading-->
<IMG WIDTH=100 HEIGHT=50
SRC="dna.gif" ALT="DNA">        <!--that image again-->
<P>                             <!--paragraph mark-->
Main Text Here!                 <!--Main Text Here!-->
<B>This point in bold</B>       <!--some bold text-->
<I>This one italic</I>          <!--some italic text-->
<P>New Paragraph                <!--another paragraph mark-->
<HR>                            <!------line across the page------->
<A HREF="http://www.ch.cam.ac.uk"> <!--An anchor, pointing to a URL-->
Click here to go to the chemistry
department home page            <!--This text is highlighted-->
</A>                            <!--End of anchor-->
</BODY>                         <!--End of body text-->
</HTML>                         <!--End of document-->
```

By convention, HTML filenames always end .html or .htm so that it is clear which files contain HTML. The first tag of the file should be <HTML> to show that the document contains HTML. With this information it is possible to write a page of hypertext. It may be easiest to begin by copying an existing HTML file and editing it. A simple example is given in Table 7.1.

The easiest way to check that the hypertext has been written correctly is simply to look at it using a web browser. Usually there will be an *Open File...* or an *Open Local...* option which will display the file. However, different browsers interpret hypertext in slightly different ways, so it is as well not to do anything too complicated, and, if possible, to check pages on different browsers.

Every document should have a useful <TITLE> to assist search programs. Titles like 'Page 1' are unhelpful. The title should give an idea of the content of the page. Every image should have an ALT text, because not all browsers can display pictures. Beware 'special' characters, such as accented letters, degree celsius symbols, Ångstroms, and so on. These may not appear as expected.

7.3.3 Chemical HTML

The WWW is particularly useful for chemists because three-dimensional information is so important for chemistry, and it can be transmitted effectively over the internet. This was emphasised in 1994 (Rzepa *et al.*, 1994), by a proposal for chemical MIME standards, which showed how it was possible to transmit molecular data electronically in a convenient way. MIME, which stands for multipurpose internet mail extensions, were intended to allow the transmission of different types of file by electronic mail, and are now used for a variety of types of electronic information exchange.

Similar ideas can be applied to the WWW, so that a suitable web-browser can display three-dimensional molecules which can be rotated, scaled and measured by the 'reader'. This opens up a new dimension for chemical publishing.

7.4 KEY POINTS

- The same computers which can calculate molecular properties can also handle databases.
- Information can be *marked-up* in some form of SGML to enable computers to search effectively.
- Computers can search for information, such as three-dimensional relationships, which cannot be held in a traditional index.
- Molecular information may be conveniently transmitted over the internet.

QUESTIONS

Q7.1 Searching the Cambridge Crystallographic Database

Does the Cambridge Crystallographic Database provide any information about boron fluoride Lewis acids coordinated to carbonyl groups (Figure 7.3)? What is the preferred conformation of such species?

This is 'any bond' not a -bond

Figure 7.3

Q7.2 Searching the Internet

Use the WWW to answer the following questions:
- What did Koopmans win a Noble prize for? (Koopmans' theorem is mentioned in Chapter 6)

- What did the composer Borodin do when not writing 'Prince Igor'?
- How many chemistry departments have WWW sites, worldwide?

Q7.3 Writing HTML

Comment, constructively, on the following HTML:

```
<HTML>
<HEAD>
<TITLE>Example</TITLE>
</HEAD>
<BODY>
<H1 ALIGN=CENTER>Is this good HTML?</H2>
<H3>A heading with some emphasis</H3>
<IMG SRC="picture.gif">
</HTML>
```

DISCUSSION

D7.1 Searching the Cambridge Crystallographic Database

The Cambridge Database contained about ten structures with a carbonyl boron fluoride complex, in 1997. In most of these, the Lewis acid is constrained within a ring, and so the preferred conformation cannot be deduced (Goodman, 1992).

D7.2 Searching the Internet

Search engines can be used to look for key words over the entire WWW. A first approach to each of the questions would be just to type in a key word and see what happens. This should work quite well for Koopmans, as his name is not common. His first name was 'Tjalling' which is unlikely to be mentioned many times on the internet. What happens if the search is done for 'Koopman', which is a common mis-spelling? An alternative approach would be to search for 'Nobel Prize' and try to find a list of the prize-winners.

A search for Borodin will mainly find references to his music, but it should be possible to discover that his main career was as a chemist.

A search for 'Chemistry Departments' is likely to produce a very large number of entries. It is probably best to look for an index of chemistry departments, rather than use a search engine, in this case. One such index is available at the Cambridge Chemistry Department (http://www.ch.cam.ac.uk/), and contained over a thousand links to chemistry departments, at the end of 1997. This is probably not comprehensive. A number of other such lists exist, all of which have different implicit definitions of 'chemistry department.'

D7.3 Writing HTML

The HTML example can be criticised in a number of ways. First, its title is very unhelpful. A search engine might find a keyword in the page, but would probably only list the title, in the first instance, and the word 'Example' does not give much information about the possible content of the page.

The first tag, < HTML >, could be made more specific by describing which version of HTML is being used. Replacing < HTML > by:

```
<!DOCTYPE HTML PUBLIC "-//IETF//DTD HTML 2.0//EN">
```

would explain that the document is version 2.0 HTML.

The first heading, H1, is terminated by an < /H2 > tag. Some browsers will allow this, but others will be confused. It is best to try not to confuse them. The second heading is < H3 > and not < H2 >. It may be that this looks neater on the browser that was used to design the document, but there is no reason to suppose it will look neater on all browsers. HTML is not ideal for producing pages with a precise appearance. Other formats, such as Adobe's PDF, are more appropriate for this. HTML should be used to reflect the structure of the document, and so < H2 > should usually follow < H1 >, unless this does not make sense in the context of the document structure.

The image does not have an ALT text, so anyone who downloads the page without the images will have no idea what information the page contains. It can also be useful to include the size of the image in the HTML.

Finally, there is no < /BODY > tag, which should occur just before the final < /HTML >.

```
<HTML>
<HEAD>
<TITLE>Example</TITLE>
</HEAD>
<BODY>
<H1 ALIGN=CENTER>Is this good HTML?</H2>
<H3>A heading with some emphasis</H3>
<IMG SRC="picture.gif">
</HTML>
```

CHAPTER 8

Applications

8.1 INTRODUCTION

Chapters one to seven have outlined the molecular modelling methods which are available. In this chapter, some applications of molecular modelling are described, in order to help to decide whether molecular modelling is likely to be useful in particular situations. Before any molecular modelling study is started, it is important to ask:

  (i) What experimental data are available?
 (ii) Is the calculation possible?
(iii) Is the calculation realistic?
(iv) What information can be obtained from the study?

Unless all four of these questions have been addressed, it may be better not to undertake a molecular modelling study. Molecular modelling can give useful quantitative data as well as inspirational qualitative information, but, for some systems, quantitative calculation may require completely unrealistic amounts of computer power.

8.1.1 What Experimental Data are Available?

Molecular modelling is not sufficiently accurate to provide reliable information about systems for which there are no experimental data. There must always be some way of checking that the calculation is making sense. Many different sorts of data may be available. Crystal structures of similar systems may give confidence that a force field is producing reasonable results. Experimentally measured selectivity may correlate with calculated results, and so increase the trust which may be put in related results for which only the computational results are available. Measurements of binding energies may be related to calculated values, taking due allowance for the uncertainties in the experimental measurements.

If, however, a molecular modelling study has no relationship to any experimental data, it can be judged only for its aesthetic value, and not for its scientific merit.

8.1.2 Is it Possible?

A molecular mechanics study can only be carried out if appropriate parameters are available. Some effects are not reproduced well by most force fields, such as π-stacking interactions, solvent effects, hydrogen bonds and hydrophobic interactions. In such cases, it is important that there should be some independent check of this part of the model.

Molecular mechanics was designed to deal with ground states of molecules. If a reaction is being studied, then it may be possible to get useful information about it by studying the starting materials and the products using molecular mechanics. If it is essential to look at the transition states, it may be possible to use molecular mechanics, and some examples are described later in the chapter.

If parameters are not available, or are insufficiently reliable, molecular orbital methods can be used. Molecular orbital methods can also be used to study transition states directly. However, they are only suitable for rather small systems, perhaps only a few first row elements even if a fairly powerful computer is available. If a study is required of a large system, for which molecular mechanics parameters are not available, it may be necessary to develop the parameters using molecular orbital theory and all the available experimental data (Appendix A.12). Alternatively, it may be possible to perform a hybrid molecular mechanics/molecular orbital theory study.

Molecular mechanics methods can cope with very large systems, but there will always be a limit for any computer. A membrane-bound protein, with an explicit solvent model and a model of the surrounding membrane, could be too large, even for molecular mechanics.

8.1.3 Is it Realistic?

Even if good parameters are available for a system, it may not be sensible to undertake a molecular modelling study. It will almost certainly be necessary to undertake a conformation search, or to investigate the conformational flexibility of the system in some other way. This may well be the time limiting step in the study. For example, a molecule such as erythromycin (Figure 8.1 and Q4.7) requires a Monte Carlo search of many thousand steps to investigate its conformation flexibility in a thorough way. If a full conformation search is to be carried out, an assessment of the time which is likely to be required should be carried out

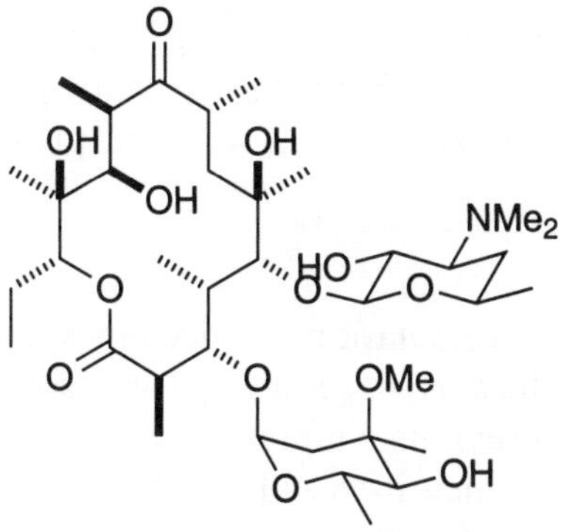

Figure 8.1 *Erythromycin A*

before it begins. If a full conformation search is not performed, there can be no great confidence that the lowest energy conformations of the molecule are being considered. In some cases, molecular dynamics may provide a more effective method of exploring conformation flexibility, particularly if experimental constraints, such as nOe data, are available. A definitive study of the folding of a protein, for which good parameters are available at an atomic level, is likely to be much too hard, unless there is experimental data, such as a crystal structure of the folded protein, to guide the calculations.

8.1.4 What Information can be Obtained from the Study?

This is the most important question of all. Some molecular modelling studies have rather vague aims, and, as a result, are likely to provide a hazy result. It is necessary to define what property is being calculated, and assess the likely accuracy of the result, so that it can be compared with an existing or a future experiment.

In order for a study to be successful, will it be enough to obtain a value for E_{MM}, the molecular mechanics energy? This may be sufficient, if the conformation of the system is the property of interest. If a value for enthalpy (H) is required, then it will probably be necessary to adjust the study so that a difference in the enthalpy (ΔH) between two closely related systems is required instead.

If a free energy (G) is required, then the study will be harder. Free energy can be calculated, but requires calculations to assess the entropy of the system. As with enthalpy, it will often be better to try to calculate differences in free energy (ΔG) rather than absolute free energies. Better still, differences of differences ($\Delta\Delta G$) may be useful quantities, for example if trends between similar systems are required. Since $G = H - TS$, if similar systems have similar entropies, then, at constant temperature: $\Delta\Delta G = \Delta\Delta H - T\Delta\Delta S \approx \Delta\Delta H$, provided $\Delta\Delta S$ is small.

8.2 STRUCTURES

In Chapter 1, Barton's calculations on decalin were described, which led to the realisation of the correct geometry for these molecules. Various of the examples in the previous chapters have looked at the structures of unbranched alkanes, which it would be natural to expect should prefer their extended conformations, but, in fact, do not. Molecular modelling has produced surprises and insights in a number of structural studies.

8.2.1 Tilivalline

Tilivalline epimerises at the C_{11} position by opening and closing the central seven membered ring (Matsumoto *et al.*, 1994, 1996). The ratio of the epimers is thermodynamically controlled, so it should be possible to calculate the proportion of each which should be formed, which is not obvious from the structure (Figure 8.2).

Matsumoto *et al.* used the MM3 force field to analyse this system. Two torsion parameters and one bending parameter were missing from the force field, and so these were determined using molecular orbital theory and checking the results against X-ray crystallographic data. A conformation search was carried out on the system, and a number of

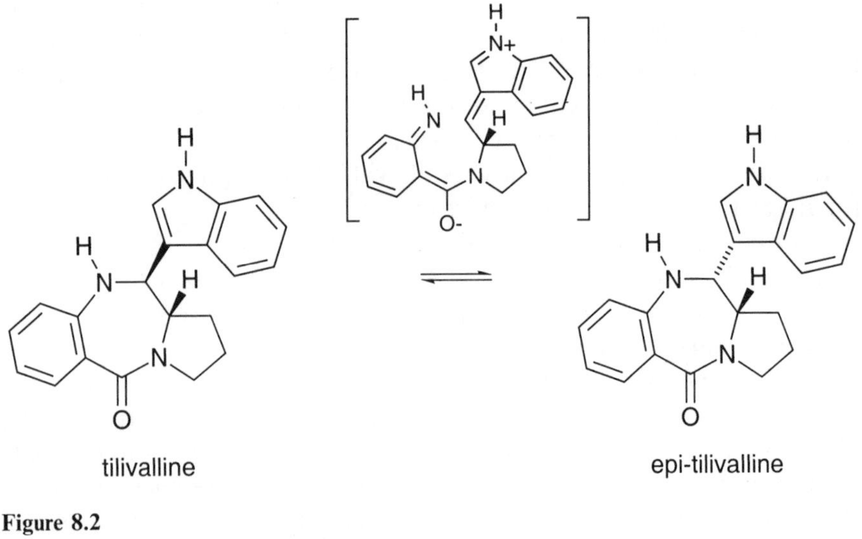

tilivalline   epi-tilivalline

Figure 8.2

conformations were found to be significantly populated at room temperature. The difference in energy between the two forms is the important quantity for this study, and the entropy of the diastereoisomers is likely to be rather similar, so it would be reasonable to use the difference in molecular mechanics energy, ΔE_{MM}, as an estimate for the difference in free energy between the diastereoisomers. Summing the Boltzmann factors for each isomer gave a result for the ratio in favour of the natural diastereoisomer and a ratio similar to that observed experimentally.

This study began with experimental data, the ratio of tilivalline to epitilivalline. The parameters for the chosen force field were not all available, but it was possible to develop the missing parameters. The conformation search that was required was not a very demanding one, and would have been completed in a reasonable time. The study asked a clear question: 'What is the ratio of diastereoisomers, assuming thermodynamic control?' and was able to provide a quantitative answer, close to the experimental result.

8.2.2 Conjugated Hydrocarbons

Elimination of a hydrogen and a good leaving group, such as chloride, to form a double bond is an important reaction (Figure 8.3), but so simple that molecular modelling should not be required to provide information about the process.

However, in more complex cases, the arrows required to eliminate hydrogen and chloride are less straightforward. Figure 8.4 shows increasingly complex examples A–D. The mechanisms of A, B, and C are fairly straightforward to draw using curly arrows to represent electron movement, but this is not true for the final example, D. The positions of the double bonds in the eliminated product are not easy to work out.

One advantage of molecular orbital calculations over molecular mechanics is that it is not necessary to define which bonds are single bonds and which are double bonds. (In fact, it is not necessary to decide where any of the bonds are.) For molecules of this size, a semi-empirical method would be most appropriate. The results are given in Table 8.1.

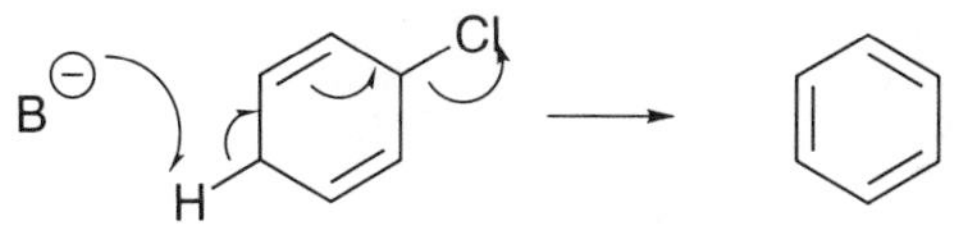

Figure 8.3

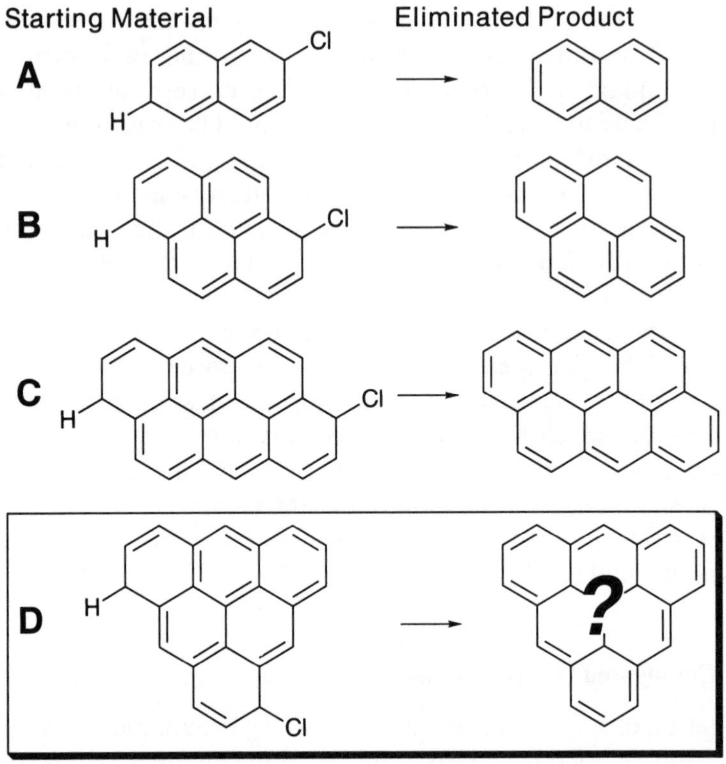

Figure 8.4

The reactions appear to be endothermic because the calculation does not take into account the energy contribution of the chloride and hydrogen ion which are removed from the system. The heats of formation of the eliminated products get larger as the molecules get larger. However, the heat of formation of D is much greater than may be expected from the rest of the molecules. C and D are the same size, and their starting materials have very similar energies, which seems reason-

Table 8.1 *MOPAC/AM1 calculations*

Elimination	Heat of formation of starting material (kJ mol^{-1})	Heat of formation of eliminated product (kJ mol^{-1})	Energy change (kJ mol^{-1})
A	172.11	169.85	−2.26
B	240.77	281.82	+41.05
C	353.53	416.15	+62.62
D	353.72	622.23	+268.51

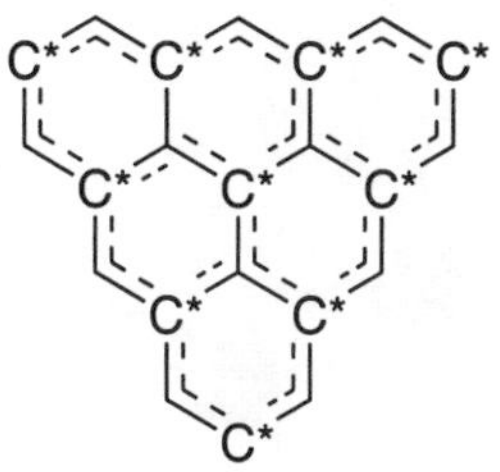

Figure 8.5

able. This makes it clear that there is something odd about the eliminated product of D. It turns out that it is not merely difficult to draw the double bonds in the product, it is impossible. This may be demonstrated by marking alternate carbons in the structure (Figure 8.5).

Every double bond must join a marked carbon (C*) to an unmarked carbon. However, there are ten marked carbons, and twelve unmarked carbons, so it is impossible to draw eleven double bonds on the structure in a chemically consistent way. It will, therefore, be much harder to do the elimination reaction D than C. With this insight it is clear that there is a problem for this particular structure, but without the insight the reassurance from molecular orbital theory that there is something odd about it may be very welcome.

The eliminated product must contain both a positive and a negative charge delocalised through the structure, or else have a triplet electronic state with two unpaired electrons. AM1 and MOPAC can also calculate the energy of the triplet state. When the energy of the eliminated products are recalculated as a triplet state, the energy of the product of D is less than the energy of C, suggesting that it is much more likely to exist as a free radical.

8.2.3 Metals

Metals are a great challenge for molecular modelling, because of the complexity of the shapes which they can adopt and the large number of ligands that they can accommodate. Molecular orbital theory can be used to analyse such structures, but the systems are necessarily rather large, and so these calculations are rather demanding. Molecular mechanics can provide useful insights into the structures of such systems.

Simply calculating the steric strain that a metal introduces into a system can be very useful in analysing metal complexation (Hancock and Martell, 1989). More sophisticated approaches have also been tried. Force fields have been designed with metals particularly in mind, for example: DREIDING (Mayo *et al.*, 1990), SHAPES (Allured *et al.*,

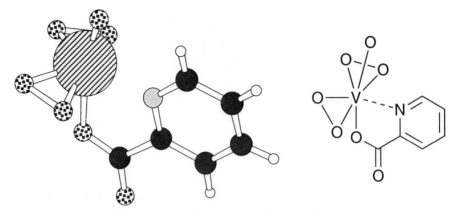

Figure 8.6 $[VO(O_2)(pic)]^{2-}$
(Shaver *et al.*, 1993)

1991) and UFF (Rappé *et al.*, 1993). A study of vanadium peroxides (Cundari *et al.*, 1997) showed that molecular mechanics could provide a rapid and accurate way of obtaining structural information about these complexes. The molecular mechanics force field was parameterised using *ab initio* calculations, as there is little experimental data for the strength of vanadium–oxygen bonds. The molecular mechanics gave results in good agreement with *ab initio* calculations, and better results than a semi-empirical (INDO) study, when compared with a number of experimental results, including $[VO(O_2)(pic)]^{2-}$ (Figure 8.6). In the case of $[V(O_2)_3F]^{2-}$ the molecular mechanics and the *ab initio* data are not consistent with the experimentally determined structure. Is this because the calculations are wrong, or because the experimental data were misleading? The calculations are not sufficiently reliable to conclude immediately that the experiment was incorrect! The data from other comparisons with experiments suggest, however, that calculations are sufficiently reliable to make it worthwhile re-examining the experimental data in the case which does not match the calculation.

8.3 REACTIVITY

Molecular mechanics is usually only used for ground state calculations, that is, minima in potential energy surfaces. Transition state calculations require the analysis of maxima or saddle points (Figure 8.7) and this cannot be done directly by most molecular mechanics programs. Transition state calculations can be done using *ab initio* molecular orbital methods, but these cannot easily cope with systems of the size that are used in synthetic chemistry. Semi-empirical methods may be insuffi-

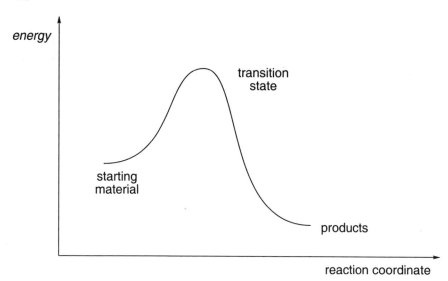

Figure 8.7 *A reaction coordinate*

ciently reliable to bridge the gap between molecular mechanics and *ab initio* methods. Molecular mechanics can, however, be used to develop useful tools for the study of transition states which can be applied to kinetically controlled reactions.

One way around the problem of using molecular mechanics to study transition states is to treat transition states as ground states. This can be done by inverting a normal mode of the system, so that the transition state is treated as a ground state (Figure 8.8). At first glance, this does not look as if it is a very good idea! There is no obvious relationship between the energy maximum at the transition state for the real reaction, and the energy minimum for the molecular mechanics model of it. However, Figure 8.8 represents an extreme case, for which there is only one degree of freedom in the system. In general, the number of degrees of freedom will be $3N-6$, where N is the number of atoms involved in the reaction, which is likely to be quite a large number for synthetic reactions. Only one of these degrees of freedom corresponds to the reaction coordinate, so inverting this component of the system leaves $3N-7$ unchanged, and these will usually account for most of the properties of the system.

In practice, it may be convenient to constrain more than one degree of freedom in the system, because the reaction coordinate may be hard to define simply in terms of a single parameter. For example, the transition state of a Cope rearrangement (Figure 8.9) may be modelled using a normal ground state force field, but with an additional term which constrains the two sp² carbon atoms which are forming a bond to be the correct distance apart, which may be found from an *ab initio* calculation.

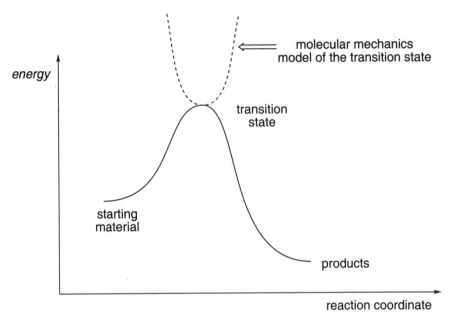

Figure 8.8 *Treating a transition state as a minimum*

Figure 8.9 *The simplest example of a Cope rearrangement.*

The model may be improved by also stretching the C—C bond which is breaking in the reaction, and perhaps by constraining bond angles and torsion angles. The values of the constraints can be derived from *ab initio* calculations and by correlation with experimental results from any Cope rearrangement which can be shown to be kinetically controlled. A more sophisticated approach would be to abandon the ground state force field for all the atoms affected by the reaction, and replacing them with entirely new parameters. The breaking and forming bonds would probably be given bond strengths weaker than normal bonds, and the bonds which change from single to double bonds would be given intermediate properties. The complete parameterisation of this sort of system is much harder than parameterisation of ground states, because it is harder to relate each parameter to a property which can be related to a mechanical model of the molecule.

Transition state modelling works best when it is used to compare similar systems, such as transition states leading to diastereomeric products. In such cases, some of the errors in the model cancel, so that the difference in energy that is calculated is much more reliable than the absolute values of energy, which may have little significance.

Transition state modelling has been pioneered by Houk (Eksterowicz and Houk, 1993; Lipkowitz and Peterson, 1993) and has been successfully applied to many reactions including hydroboration, radical additions, pericyclic reactions, aldol reactions and nucleophilic additions to carbonyls. The significance of the process has been questioned, however. Menger and Sherrod (Sherrod, 1990; Menger and Sherrod, 1990) demonstrated for acid-catalysed lactonisation that it is possible to get better correlations to experimental results by using 'transition state models' which were unrelated to the transition states! All the experimental data that were available were used to parameterise these models, and they were not applied to systems which had not been used in their parameterisation. If a transition state force field is to be useful, it must be applicable to systems which were not used in the creation of the model. In such cases, ensuring that the model is as close to chemical understanding as possible should help the predictive power of the model. The conceptual problem is close to that of applying molecular mechanics calculations to ground states. A force field must be tested with all the experimental data that are available, in order to have confidence in its ability to give useful information about systems not used in the parameterisation process. The difference with transition state studies is that the amount of data available for parameterisation and testing is usually rather small, and there are only very rarely direct experimental data on the transition state itself. A transition state model is only of value if it can be useful in the understanding of reactions and in the design of improved procedures. In some cases, this has been demonstrated.

8.3.1 Diels–Alder Reactions

A model of the Diels–Alder reaction was first produced by Brown and Houk (1985) and was based on Allinger's MM2 force field. This model was used by Takahashi in a synthesis of the steroid skeleton (Takahashi *et al.*, 1988). The starting material for the key reaction, illustrated in Figure 8.10, can undergo a Diels–Alder reaction to form any of the four products, **A**, **B**, **C** or **D**. It is obvious neither by inspection nor by model building which of the four will be preferred. The force-field calculation gave the result that **A** had the lowest energy, and was almost $14\,\text{kJ}\,\text{mol}^{-1}$ lower in energy than the second lowest (**C**). This suggested that **A** should be the main product, and that less than 2% of the competing products should form at the reaction temperature of 180 °C. The only product that was detected was **A**.

In a related study (Figure 8.11, Takahashi *et al.*, 1992), the product

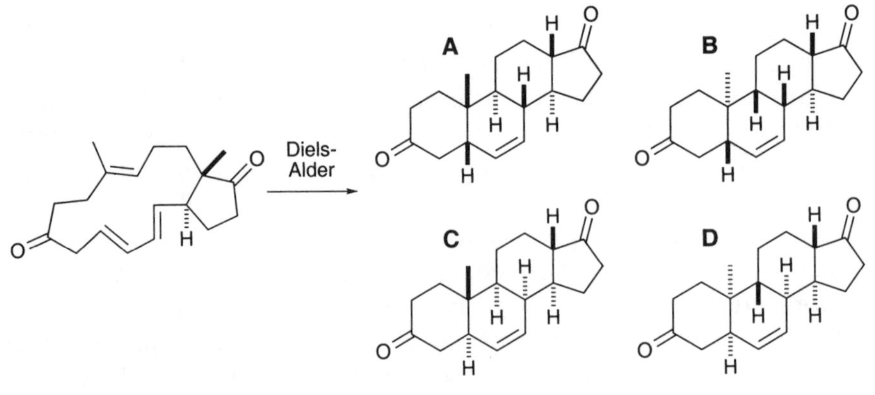

Figure 8.10 *Which Diels–Alder product will form?*

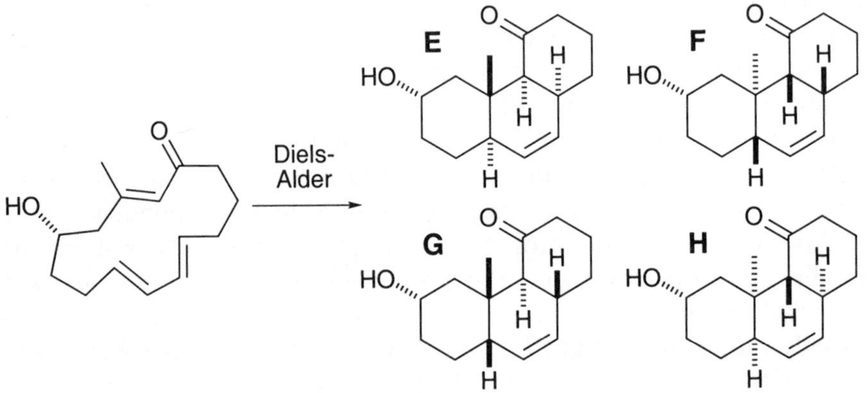

Figure 8.11 *Calculated and experimental ratios in close agreement*

ratio of a Diels–Alder reaction was calculated to be 94 : 2 : 4 : 0 and found experimentally to be 93 : 0 : 7 : 0 for **E** : **F** : **G** : **H**. A single chiral centre creates four new chiral centres in a selective and predictable way. These results give confidence that the molecular mechanics model of the Diels–Alder reaction works well and can be used in a predictive way, for similar systems.

The question of what constitutes a similar system is a very difficult one. Will the model only work for fourteen-membered rings? What about acyclic systems? What about intermolecular reactions? What about Lewis-acid catalysis? Confidence in the method must diminish, as it is applied to systems less similar to those for which the method was developed. A study of Lewis-acid catalysed intramolecular Diels–Alder reactions developed a revised force field which appears to work well for these systems (Raimondi *et al.*, 1992). This study notes that a different force field is required for the activated and the unactivated systems.

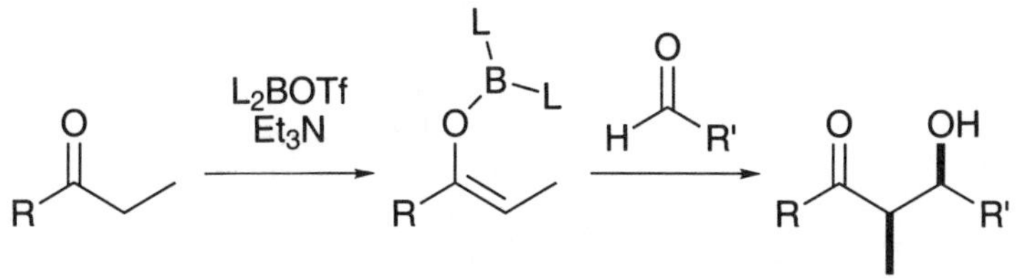

Figure 8.12 *The boron-mediated aldol reaction*

8.3.2 Boron-mediated Aldol Reactions

The boron-mediated aldol reaction (Figure 8.12) is an important carbon–carbon bond forming reaction which forms two new chiral centres. With the conditions illustrated, the two new chiral centres are *syn* to each other. If a chiral ketone or a chiral aldehyde is used in the reaction, the chirality of the substrates may be transferred to the product. Alternatively, chiral ligands, L, can be put on the boron to introduce absolute selectivity into the reaction. It is not obvious, however, how these chiral effects would be transmitted. If the ketone and aldehyde were both chiral, would they work together or would their effects oppose each other (double asymmetric induction)?

The first step of the reaction is the treatment of the ketone with a dialkyl boron triflate and a tertiary amine base. This forms an enol borinate. What conformation will this adopt? An *ab initio* molecular orbital study was undertaken of these species in order to find out (Goodman *et al.*, 1990). This showed that the oxygen–boron bond would not be in the same plane as the carbon–carbon double bond, but twisted out of the plane (Figure 8.13). With hindsight, this should probably have been expected, but the calculation provided this piece of information without the benefit of reading the literature on similar compounds. The *ab initio* calculations were then used to build a force field for enol-borinates.

The next stage of the study was to develop a force field for the transition state of the aldol reaction (Bernardi *et al.*, 1990), building on an earlier *ab initio* study of the transition state (Li *et al.*, 1988). The transition state is thought to be a six-membered ring (the Zimmerman–Traxler transition state, Zimmerman and Traxler, 1957), which is illustrated in Figure 8.14. As well as the chair conformation, boat conformations may also be possible.

Ab initio calculations on the transition state produced another surprise. There were three possible conformations for the transition state: a chair and two different boats (Figure 8.15). Searching for transition states is very difficult, compared with searching for minima, because simply 'walking downhill' will lead to a minimum and not a

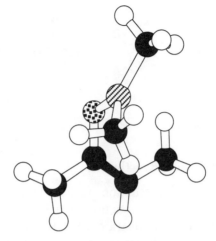

Figure 8.13 *The non-planar structure of an enol borinate*

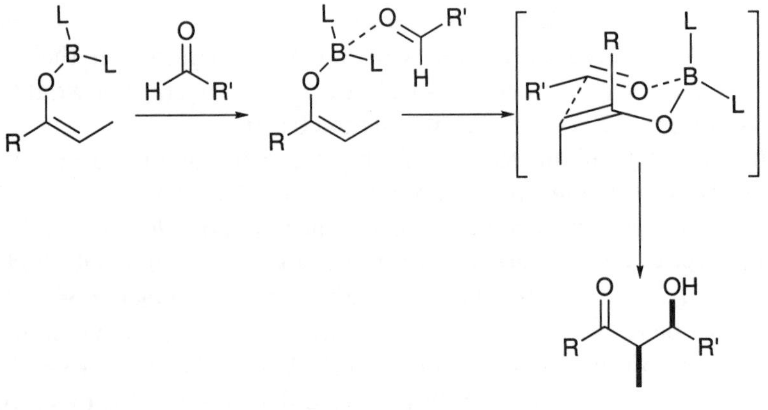

Figure 8.14 *The aldol reaction, going through a chair transition state.*

transition state. Algorithms exist to find transition states, but they will usually only work if the starting point is very close to a transition state (Baker, 1986). As a result, there is no absolute confidence that all possible transition states for the carbon–carbon bond forming reaction have been found.

The three transition states were calculated for the reaction of formaldehyde with ethanal, with hydrogen ligands on boron, because the computers available could not cope with *ab initio* calculations on much larger systems. The calculations were repeated for all the analogues with one methyl group, substituting each hydrogen in turn. Finding these transition states was relatively easy, because it was possible to take the all-hydrogen transition state, exchange one hydrogen for a methyl

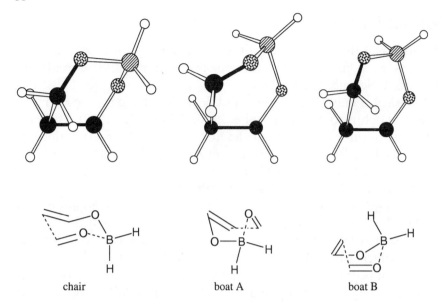

chair   boat A   boat B

Figure 8.15 *The three transition states for the boron-mediated aldol reaction of formaldehyde with ethanal*

group, and then minimise only those bond lengths and angles which were associated with this new group. The resulting structure was usually very close to a transition state, and so a transition-state finding algorithm would automatically adjust the geometry slightly to reach the transition state. These calculations were used to create a force field for the transition state.

The force field was used to analyse the stereoselectivity of the reaction (Bernardi *et al.*, 1995) and proved to be useful. In addition to rationalising existing results, it was used to design a new ligand for boron in this reaction (Gennari *et al.*, 1992), which worked very effectively.

8.3.3 Conclusions

These studies, and other related studies, show that transition state modelling can play a useful role in the analysis of stereoselectivity in organic reactions. The same principles apply to modelling transition states as to modelling ground states: as much data as possible must be used to develop and validate force fields, and the model studies should not be allowed to move too far from experimental data. The problems with modelling transition states are that the amount of data available is likely to be small, and correlations with experiment are usually indirect. As a result, the data from such studies must be interpreted with caution.

8.4 DRUG DESIGN

In principle, the development of a new drug could follow a three-step process. First, an enzyme or receptor involved in a disease needs to be identified. Second, the structure of the enzyme or receptor needs to be established. Finally, a molecule must be designed and synthesised which will bind tightly to the enzyme or receptor (Whitesides, 1990). There is no shortage of information about the first step. Good structural information is lacking for many interesting enzymes and receptors, but a huge amount of data is available. The genes for many enzymes and receptors have been cloned, so making it possible to obtain the compounds in sufficient quantity to find their structures by X-ray crystallography or by NMR. The problem is the third step. Even with a good three-dimensional structure for an enzyme, it is not easy to design and synthesise a ligand which will bind tightly and specifically to it. Such a molecule would not necessarily be a good drug, of course, because it would need further testing for toxicity and side effects, and a large scale synthesis would need to be designed. Finding a good tight-binding ligand is, however, a major problem, and a rapid, reliable method of doing this would revolutionise the development of new drugs.

The earlier chapters have described how it is possible to calculate structures quite reliably. In order to inhibit an enzyme or block a receptor, all that needs to be done is to design a ligand which binds strongly to it. This ought to be possible using the techniques discussed so far (Cohen *et al.*, 1990). It may even be possible to calculate how a molecule will react in an enzyme active site! So why has the problem of ligand design not been satisfactorily solved?

A number of factors combine to make this problem an extremely difficult one:

- Will a particular molecule fit in an active site?
- What holds it in?
- How tightly do these ligands bind?
- How can a different molecule be designed which will also fit?

8.4.1 Does a Molecule Fit into an Active Site?

For a molecule to fit in an active site at all, it must not be too large or the wrong shape. Molecular mechanics can calculate this rather well, assuming that the structure of the enzyme is rigid. This will not, of course, be the case. If an inhibitor does not seem to fit in the active site, as determined by X-ray crystallography, it may be that it can just push a small part of the structure aside, or displace a water molecule, to bind very tightly. The calculation of whether or not a molecule can fit in the active site is not trivial, therefore.

8.4.2 What Holds Molecules in Active Sites?

The forces that hold a molecule in an enzyme active site or in a receptor are precisely the ones which molecular mechanics does not calculate very well. Hydrogen bonds may well be important inter-actions, and bonds to metals. The hydrophobic effect may also be significant. Non-polar groups, such as alkyl chains or aromatic rings, prefer to be close to other non-polar groups, rather than next to water. In order to calculate the size of this effect, it is necessary to have some idea of how the molecule interacts with water, and this is a difficult calculation. The interactions between non-polar groups may also be complicated. Aromatic rings interact favourably, provided they are in the right orientation with respect to each other (Hunter and Sanders, 1990). This sort of detail is often calculated rather imprecisely by many force fields.

Ligands may be held in active sites in a counter-intuitive way. For example, dihydrofolate and methotrexate (Figure 8.16) both bind to the same active site. The side chain, R, is identical for both molecules. The structures look rather similar, and so it would be reasonable to expect that they have the relative orientation illustrated in the figure.

In fact, the illustrated part of methotrexate binds to the active site upside down, relative to dihydrofolate (Bolin *et al.*, 1982)! The approx-imate relative orientations of the bound form are illustrated in Figure 8.17. It may seem that the molecules are less well aligned in this reversed arrangement, but this is not the case. The reason for this is found in the electrostatic properties of the molecules, which are illustrated in Plate 6. The blue region of both molecules corresponds to the most negatively charged area, which is most likely to interact with a positive charge or a

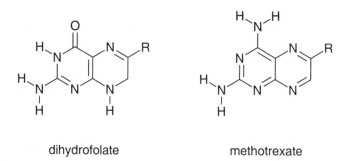

dihydrofolate   methotrexate

Figure 8.16 *Dihydrofolate and methotrexate*

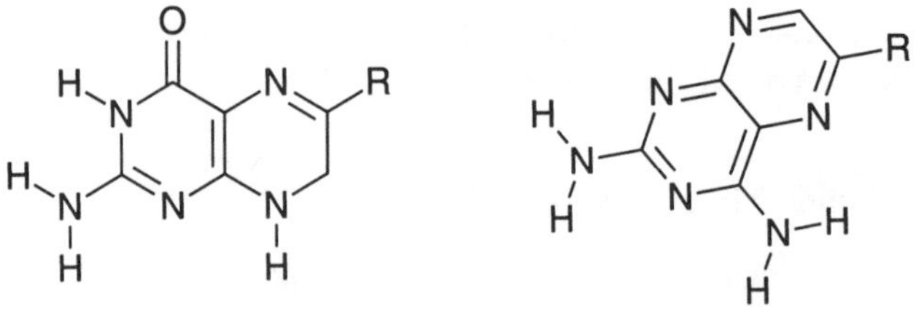

dihydrofolate   methotrexate

Figure 8.17 *The relative orientations of bound dihydrofolate and methotrexate*

hydrogen bond donor, and is on the opposite side of the two systems, when drawn in the orientation of Figure 8.16. The electrostatics match much better by inverting one of the molecules.

8.4.3 How Tightly do Ligands Bind?

Calculation of binding energy is made harder still because of problems in calculating the energy of the unbound state. The binding site may well contain water molecules which will be released on binding to a ligand, causing a significant entropy change. The unbound ligand will be surrounded by water molecules and interacting with them in some way. This is a situation where the enthalpy change (ΔH) for the binding process may be quite different to the free energy difference (ΔG), and so simply calculating the molecular mechanics energies (E_{MM}) of the bound and unbound states is unlikely to give reliable answers. This problem may be partially circumvented by calculating the relative binding affinity of structurally similar ligands, rather than trying to calculate absolute binding affinities. Such a limited study may be useful, but it would be better to be able to do a fuller calculation. The experiments which measure the binding affinities naturally have a level of uncertainty which may be hard to quantify.

8.4.4 How Can a Molecule be Designed to Fit in an Active Site?

Even if good structural data about an active site are available, and a molecular mechanics model of the interactions which control binding is trusted for this case, it is still very hard to generate a new lead, a novel molecule which will bind well (Dean 1987; Kuntz, 1992; Böhm and Klebe, 1996).

One way to address the problem is to use structures from a structural database and to try to dock each one into an active site. This requires a rapid method of fitting and evaluating the binding of each trial structure found. A widely used program which does this is called DOCK (Kuntz *et al.*, 1982). The fitting and evaluating process is time consuming, if a large database of compounds is considered, but it only considers the conformation of each molecule which is present in the database. It would be better to consider all low energy conformations of all the molecules in the database, except that this would substantially increase the complexity of the calculation, perhaps to well beyond the point for which the cost of the calculation is affordable. The original version of DOCK only considered single conformations of potential ligands, but the increase in available computer power has led to the development of algorithms for flexible docking (Makino and Kuntz, 1997).

A widely used program which can pick out interesting parts of an active site is called GRID (Goodford, 1985). This uses a variety of small probe groups, including methyl, hydroxyl and carbonyl-oxygen, which are used to discover which parts of the active site give favourable interactions. These data can be used to suggest desirable relative orientations of methyl groups, hydroxyl groups, carbonyls and so on, within a potential ligand.

Molecules which interact with the same receptor or enzyme often have features in common, such as an aromatic ring a particular distance from a carboxylate. By comparing a series of active molecules, it may be possible to develop a pharmacophore, that is, a particular arrangement of functional groups which the active molecules have in common. Pharmacophore hypotheses are used to design new, potentially active compounds. If the structure of the active site is know, an alternative approach is possible. The active site can be characterised by picking out the interesting features, for example a hydrophobic pocket which neatly fits a methyl group, or a hydrogen bond acceptor, or a metal which would interact well with a carboxylate. This can lead to a three-dimensional pharmacophore.

An approach to the design of ligands is to build up suitable molecules inside the active site (Lewis and Leach, 1994). A number of programs are available to do this, including LUDI (Böhm, 1992, 1994), SPROUT (Gillet *et al.*, 1995) and programs by Dean (Dean *et al.*, 1995). Starting from a description of an active site, potential ligand molecules are built up or formed from fragments. CAVEAT (Lauri and Bartlett, 1994) uses a different approach: a database is searched for molecules with functional groups in particular orientations.

Using these ideas, it is possible to obtain structures which may bind well to an active site. However, it may not be possible to make the structures, or they may have other undesirable properties. These techniques are useful for lead generation, but the optimisation of the leads is still the preserve of the medicinal chemist.

8.4.5 Determining Protein Structures

Molecular modelling makes a contribution to another part of the process of drug design: structure determination. X-ray crystallography of proteins gives an electron density map, which is not readily interpreted as a series of atoms. The structure which is finally deposited in the Brookhaven database is produced by fitting a molecular structure to the available data. In a similar way, NMR determination of protein structures depends on a force fields optimisation of the structure to fit the experimental data, using a program such as Xplor (Brünger, 1990; Nilges *et al.*, 1991).

It is tempting to take this process one stage further. There is much more DNA sequence data available than there are data on the three-dimensional structures of enzymes and receptors. The Human Genome Project will provide a complete description of the DNA of humans, including every enzyme and receptor. In many cases, the sequence data contain the information that the protein needs to fold up into its preferred three-dimensional shape. Might it be possible to use these data to predict the three-dimensional structure of proteins, without the need for complex and expensive X-ray crystallography or NMR spectrometry?

In principle, molecular mechanics may be able to do this, using long molecular dynamics to simulate the folding of a protein. Unfortunately, the calculations would be so hard that it is difficult to foresee a time when they may become possible, using atomic-scale models of the molecules. It is possible to simulate the unfolding of proteins, starting from an experimentally determined three-dimensional structure. For example, Jorgensen has simulated the unfolding of barnase in urea (Tirado-Rives *et al.*, 1997). The molecular dynamics simulations were run for several nanoseconds (which corresponds to millions of steps) and produced results consistent with experimental data. The calculation suggests that urea denatures the protein by stabilising the unfolded form rather than destabilising the folded form. Calculating how the protein folded to its native state from a denatured state is not worth attempting, because it would be so much harder than this calculation of unfolding, which currently represents the limit of the possible.

8.5 CONCLUSIONS

Molecular modelling can provide useful information about a wide range of chemical problems. The precision of the results diminishes with the increasing complexity or the systems being studied, so the technique is very good for structural studies, must be treated with more caution for reactivity investigations, and may provide inspiration as well as some quantitative data for drug design

Some of the properties of molecules are too complex to be analysed without computers. An important example of this is the conformational analysis of flexible molecules. Conformation searching is usually too hard for humans, but it is possible for computers. As a result, computers can produce ideas, or at least can help people to have ideas by high-lighting the consequences of simple descriptions of structure applied to complicated systems. Molecular mechanics can be misused (Lipkowitz, 1995), but it is extremely powerful if used carefully, and if the problems that are chosen are carefully defined. Molecular modelling can be very useful in synthesis, drug design, medicinal chemistry, materials design, mechanisms, inorganic mechanisms and synthesis, biological chemistry and other areas of molecular sciences. The cost of the technique is diminishing as computers become cheaper and more powerful, and programs use the increased power to become more accessible to the non-expert as well as to perform more sophisticated calculations. Molecular modelling is a tool which can both assist and inspire the experimental chemist and so its use should not be confined to specialists.

Glossary

3-21G, 6-31G**: Examples of Gaussian Basis sets, using Pople's nomenclature

AM1: Austin Model 1. A semi-empirical molecular orbital method (Chapter 6)

AMBER: A molecular mechanics force field (Appendix A.7)

BDNR: Block Diagonal Newton Raphson. A minimisation algorithm

BFGS: Broyden–Fletcher–Goldfarb–Shanno. A minimisation algorithm

BLYP: A functional for density functional calculations

CADPAC: Cambridge Analytical Derivatives Package. An *ab initio* molecular orbital theory package (Chapter 6)

CCD: Cambridge Crystallographic Database

CFF93: A molecular mechanics force field (Appendix A.7)

CHARMm: A molecular mechanics force field (Appendix A.7)

CI: Configuration Interaction. A technique in *ab initio* molecular orbital theory

CML: Chemical Markup Language; a SGML for chemistry, designed by Peter Murray-Rust

CNDO: Complete Neglect of Differential Overlap. A semi-empirical molecular orbital method (Chapter 6)

COSMIC: A molecular mechanics force field (Appendix A.7), and also a molecular modelling program

CPK: Corey, Pauling and Kulton design for plastic models of molecules

CPU: Central Processing Unit. The 'brain' of a computer

CVFF: Consistent Valence Force Field. See CFF93

DFT: Density Functional Theory. A new approach to molecular orbital theory

DNA: Deoxyribonucleic acid

DREIDING: A molecular mechanics force field (Appendix A.7)

dtd: Document Type Definition. This is an explanation of all the labels that may be used in a SGML

E_{MM}: Molecular mechanics energy

ECEPP: A molecular mechanics force field (Appendix A.7)

ESP: Electrostatic Potential. ESP-derived atom-centred point charges are based on the electrostatic field around a molecule

FEP: Free Energy Perturbation

GAMESS: An *ab initio* molecular orbital theory package (Chapter 6)

Gaussian: An *ab initio* molecular orbital theory package (Chapter 6)

Global minimum: The lowest energy point on a potential energy surface

GROMOS: A molecular mechanics force field (Appendix A.7)

Hessian: A matrix of the second derivatives of energy with respect to molecular coordinates. The Hessian can be used to determine whether a stationary point is a minimum, a transition state with one negative normal mode or a higher order saddle point

HOMO: Highest Occupied Molecular Orbital

HTML: HyperText Markup Language. This is the language that WWW browsers understand

http: HyperText Transfer Protocol

Java: A computer language which was designed with World Wide Web applications particularly in mind

Javascript: A scripting language for World Wide Web browsers, which is not closely related to Java

LCAO: Linear Combinations of Atomic Orbitals. A technique used to build up molecular orbitals

Local minimum: A structure that is minimised with respect to all its coordinates, but higher in energy than the global minimum

LUMO: Lowest Unoccupied Molecular Orbital

MacroModel: A molecular modelling program (Mohamedi *et al.*, 1990)

MC: Monte Carlo. The name for any of a wide range of stochastic methods which involve random numbers, or even a mountain with a casino. The meaning may be clear from the context

MD: Molecular Dynamics

MIME: Multipurpose Internet Mail Extension. A MIME-type describes the sort of information that a mail message, or other computer file, contains, and so a computer knows whether to expect an image, a molecule or a spectrum, for example

MINDO/3: A semi-empirical molecular orbital method (Chapter 6)

MM2, MM3, MM4: Molecular mechanics force fields (Appendix A.7)

MM2*: MM2 as implemented in MacroModel

MMFF: A molecular mechanics force field (Appendix A.7)

MNDO: A semi-empirical molecular orbital method (Chapter 6)

MOPAC: A semi-empirical molecular orbital program (Chapter 6)

MP2: Second-order Møller–Plesset correction to a Hartree–Fock calculation

Multiplicity: A measure of the number of unpaired electrons in a molecule. Singlet multiplicity means that all the electron spins are paired, a doublet must have one unpaired spin

node: A stationary point in a standing wave. In the context of molecular orbital theory, this is a point at which the wave function is zero. For example, the $2p_x$ orbital has the nodal plane $x = 0$

nOe: Nuclear Overhauser Effect. Used in NMR spectroscopy to determine which atoms are close to each other

OPLS: A molecular mechanics force field (Appendix A.7)

PDB: Brookhaven Protein Data Bank

PDF: Portable Document Format

PM3: Parameterised Model 3. A Semi-empirical molecular orbital theory Hamiltonian, developed by Stewart

PRCG: Polak–Ribiere Conjugate Gradient algorithm for minimisation

QSAR: Quantitative Structure Activity Relationship

RHF: Restricted Hartree–Fock. Useful approximation in *ab initio* molecular orbital theory, forcing all electrons to be paired

SCF: Self Consistent Field. Use in molecular orbital theory

SGML: Standard Generalised Markup Language. HTML is a SGML with a particular dtd

SHAKE: A method of speeding up molecular dynamics simulations by constraining C—H bond lengths

Simplex: A simple, minimisation algorithm which does not require the calculation of derivatives

S_N1: Substitution, Nucleophilic, Unimolecular

S_N2 : Substitution, Nucleophilic, Bimolecular

SOMO: Semi-occupied Molecular Orbital – Used instead of HOMO or LUMO when the highest occupied orbital contains only one electron.

Spartan: An *ab initio* molecular orbital theory package (Chapter 6)

STO: Slater Type Atomic Orbital: An early basis set for molecular orbital theory. These are close in shape to atomic orbitals, but much harder to manipulate mathematically than Gaussian functions, so the latter are now used almost exclusively

TIP3P, TIP4P: Models for the properties of water molecules (Jorgensen *et al.*, 1983)

Tripos: A molecular mechanics force field (Appendix A.7)

UFF: A molecular mechanics force field (Appendix A.7)

UHF: Unrestricted Hartee–Fock. Unlike RHF, this permits a system to have any multiplicity. Can lead to spin contamination

URL: Uniform Resource Locator—the address of a WWW page

WWW: World Wide Web

XED: Extended Electron Distribution (Vinter, 1994).

Z-matrix: See Appendix A.8

Appendices

A.1 PHYSICAL CONSTANTS AND CONVERSION FACTORS

Constants:

$R = kN_A = 8.314 \text{ J K}^{-1} \text{ mol}^{-1}$	Gas Constant
$N_A = 6.022 \times 10^{23} \text{ mol}^{-1}$	Avogadro's Number
$h = 6.6261 \times 10^{-34} \text{ J s}$	Planck's Constant
$k = 1.3806 \times 10^{-23} \text{ J K}^{-1}$	Boltzmann Constant
$e = 1.60218 \times 10^{-19} \text{ C}$	Electronic Charge
$c = 2.9979 \times 10^8 \text{ m s}^{-1}$	Speed of light in a vacuum
$\varepsilon_0 = 8.854 \times 10^{-12} \text{ C}^2 \text{ J}^{-1} \text{ m}^{-1}$	Permittivity of a vacuum
$a_0 = 5.2918 \times 10^{-11} \text{ m}$	Bohr Radius

Conversion Factors:

$273.15 \text{ K} = 0\,°\text{C}$
$1 \text{ Torr} = 1 \text{ mm Hg}$
$760 \text{ mm Hg} = 1 \text{ bar at } 298 \text{ K}$
$1 \text{ atmosphere} = 101\,325 \text{ Pa}$
$1 \text{ bar} = 100\,000 \text{ Pa}$
$1 \text{ mol of gas occupies } 22.4 \text{ dm}^3$
$1 \text{ dyne} = 10^{-5} \text{ N}$
$1 \text{ mdyn Å}^{-1} (\text{atom})^{-1} = 602 \text{ kJ mol}^{-1} \text{ Å}^{-2}$
$RT \approx 2.5 \text{ kJ mol}^{-1}$ at room temperature

Units of Energy

	Hartree	eV	cm^{-1}	kcal mol^{-1}	kJ mol^{-1}
Hartree	1	27.211	219 475	627.52	2626
eV	0.03675	1	8066	23.06	96.48
cm^{-1}	4.556×10^{-6}	0.000 123 9	1	0.002 86	0.012
kcal mol^{-1}	0.001 593 5	0.043 365	349.65	1	4.184
kJ mol^{-1}	0.000 380 8	0.103 65	83.57	0.239	1

A.2 ELECTROMAGNETIC SPECTRUM

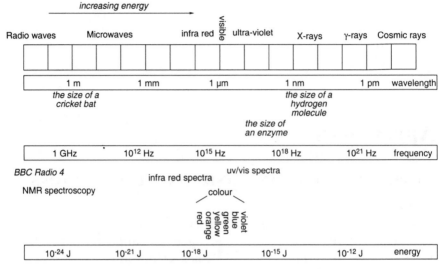

A.3 BOLTZMANN FACTORS

Boltzmann factors are a simple way of relating energy differences to population differences. Consider two structures, A and B, with energies E_A and E_B, expressed in J mol^{-1}. The Boltzmann factors are $\exp(-E_A/RT)$ and $\exp(-E_B/RT)$, respectively, where T is the temperature measured in Kelvin, and R is the gas constant (8.314 J K^{-1} mol^{-1}). There is a danger of confusion over units, because molecular mechanics energies are usually expressed in *kilo*Joules per mole, and not in Joules per mole. The ratio of populations will be the ratio of Boltzmann factors. Thus, if the energies are equal, the populations should also be equal. If the energies differ by 1 kJ mol^{-1} at 25 °C, the population ratio will be 1.5 : 1 in favour of the lower energy species.

To calculate the relative populations of families of molecules, the Boltzmann factor of each member of each family must be calculated individually. The relative populations of the two groups will be the ratio of the sum of the Boltzmann factors.

The results of Boltzmann factor calculations for two structures may be expressed graphically (Figure A.1).

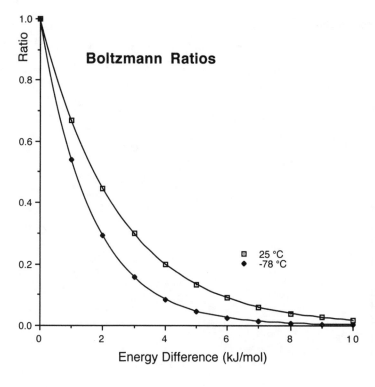

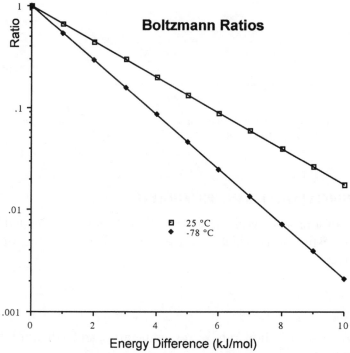

Figure A.1 *Boltzmann ratios*

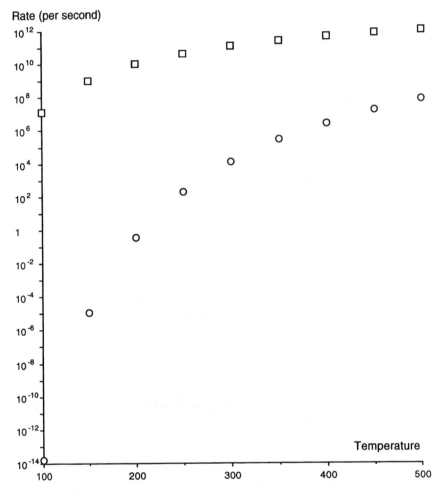

Figure A.2 *Variation of rate with temperature, for different activation energies;*
$(\square)\ \Delta G^{\ddagger} = 10\,\text{kJ}\,\text{mol}^{-1},\ (\bigcirc)\ \Delta G^{\ddagger} = 50\,\text{kJ}\,\text{mol}^{-1}$

A.4 APPROXIMATE KINETIC GRAPH

The rate of a process with energy barrier $\Delta G^{\ddagger}$ may be estimated by the following expression, which gives a rough figure for the order of magnitude:

$$\frac{kT}{h}\exp\left(\frac{-\Delta G^{\ddagger}}{RT}\right)$$

Figure A.2 and A.3 show rate *versus* temperature and activation energy, respectively.

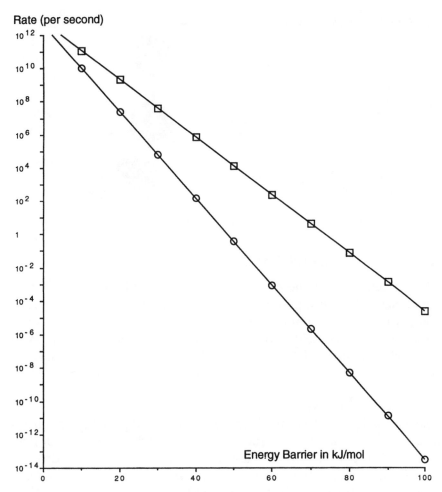

Figure A.3 *Variation of rate with activation energy for different temperatures;*
(□) T = 300 K, *(○) T =* 200 K

A.5 VAN DER WAALS RADII

H 1.06

B 1.65	C 1.53	N 1.46	O 1.42	F 1.40
	Si 1.93	P 1.86	S 1.80	Cl 1.75
	Ge 1.98	As 1.94	Se 1.90	Br 1.87
	Sn 2.16	Sb 2.12	Te 2.08	I 2.04

All figures in Ångstrom (Bondi, 1964)

A.6 STEREOCHEMICAL ASSIGNMENT

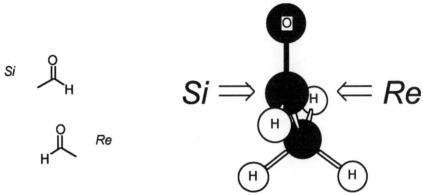

A.7 A LIST OF FORCE FIELDS

The references given here are intended as a lead, rather than as a comprehensive description of each force field. Most force fields are so complicated that it is not possible to describe them in detail with a single paper. Some force field authors maintain URLs for the World Wide Web, which point to the latest information about the force fields. A few of these are included, which were tested at the time of writing.

AMBER: 'Assisted Model Building with Energy Refinement'. This force field was designed by Kollman for the simulation of peptides and nucleic acids (Weiner *et al.*, 1984; Cornell *et al.*, 1995). It is widely used for these systems. It is parameterised to use either implicit or explicit hydrogens. The former makes the calculations rapid, at the expense of lesser accuracy. http://www.amber.ucsf.edu/amber/

CHARMm: 'Chemistry at Harvard Macromolecular Mechanics'. CHARMm is the name both for a force field and for the program which manipulates it. They were designed by Karplus to model macro-molecular structures, and proteins in particular (Brooks, 1983; Blondel and Karplus, 1996).

CFF93: This force field was designed by Hagler for accurate definitions of both small and large molecules, and is designated a 'Class II' force field by its creators because it contains parameters for anharmonicity and coupling between different distortions. This approach leads to higher accuracy at the expense of greater complexity, and a very large number of parameters are required for the force field. In addition, seven scaling parameters are introduced. A large number of parameters should lead to a concern about their transferability, and this issue is carefully addressed by the authors (Hwang *et al.*, 1994). The force field has its roots in the earlier consistent force field, CFF (Lifson *et al.*, 1979).

COSMIC 'COmputation and Structure Manipulation In Chemistry' (Vinter *et al.*, 1987; Morley *et al.*, 1991). COSMIC is the name both of a force field and a molecular mechanics package designed to manipulate it, and was designed to be a general purpose force field.

DREIDING: A simple generic force field for predicting the structures and dynamics of organic, biological and main-group inorganic molecules (Mayo *et al.*, 1990).

ECEPP: 'Empirical Conformational Energy Program for Peptides'. Developed by Scheraga (Momany *et al.*, 1975).

GROMOS: 'Groningen Molecular Simulation'. A general molecular mechanics force field (van Gunsteren and Berendsen, 1990). http://igc.ethz.ch/gromos/welcome.html

MM2: Allinger's MM2 force field, and its various derivatives, were designed for small molecules and give good results for a wide range of systems (Allinger, 1977). This force field is very widely used, and may be regarded as setting the standard against which other force fields should be judged.

MM3: (Allinger *et al.*, 1989). This force field corrects some of the weaknesses of MM2, particularly concentrating on vibrational frequencies, which MM2 does not reproduce very well. Whilst it is similar to MM2 in many respects, new parameters were added, which increased the complexity and accuracy for the systems for which the new force field was designed.

MM4: (Allinger *et al.*, 1996). A further increase in complexity and accuracy over MM3, concentrating on vibrational frequencies and rotational barriers, which are particularly difficult to calculate precisely.

MMFF: 'Merck Molecular Force Field' (Halgren, 1996). This force field seeks to achieve high accuracy for small molecules and also for large proteins. It is based on high level *ab initio* calculations, and is particularly suited to molecular dynamics.

OPLS: 'Optimised Potentials for Liquid Simulations'. Designed to model proteins in solution and is compatible with TIP4P, TIP3P and SPC models for water (Jorgensen and Tirado-Rives, 1988).

SHAPES: An empirical force field designed particularly for transition metal complexes (Allured *et al.* 1991).

Tripos: The force field implemented in the Sybyl package, from Tripos Associates (Clark *et al.*, 1989). This force field has evolved from COSMIC, and the force field of White (White and Bovill, 1977).

UFF: 'Universal Force Field'. The parameters for this force field are based only on the elements, and not on groups of atoms as is usual (Rappé *et al.*, 1992, 1993). This approach should make it possible to provide a force field which can calculate energies for any molecule, not just a limited range of organic systems. This can be very useful, but it may be harder to assess how reliable the force field will be in a particular situation.

A.8 Z-MATRICES

It is sometimes convenient to describe a molecule in terms of internal coordinates, using a Z-matrix, rather than using Cartesian Coordinates. This means that the position of each atom is expressed in terms of the positions of atoms which have already been defined. Thus a typical line in a Z-matrix description of a molecule looks like:

AtomType r *atom 1* θ *atom 2* ϕ *atom 3*

The first item is the sort of atom that is being described. The second, r, is the distance from this new atom to another atom *atom 1*. This atom must have already been described earlier in the Z-matrix. There is now an angle, θ, which is the angle created by the new atom, *atom 1* and *atom 2*. A second angle, ϕ, describes the torsion angle between the new atom and *atoms 1–3* (Figure A.4).

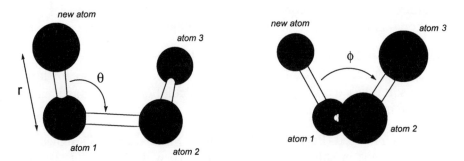

Figure A.4

This description depends on there being three atoms that are already defined, so the beginning of the Z-matrix is slightly different. The first atom is usually just given an atom type, and no information about its position. The second atom will be defined simply by its distance from the first atom. The third by its angle. For example, hydrogen peroxide (HOOH), is a simple four-atom molecule whose Z-matrix (minimised by AM1) is as follows:

```
H1
O1   1.1   H1
O2   1.5   O1   107.0   H1
H2   1.1   O2   107.0   O1   100.0   H1
```

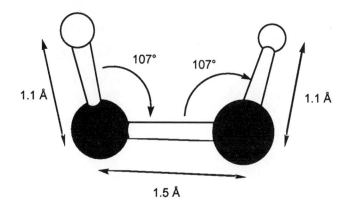

An internal coordinate description of a molecule has the feature that it is not necessary to define a position or an orientation for the molecule, since all of the atoms are defined relative to each other. It is also possible to rotate around torsion angles, by making simple changes to the Z-matrix. For example, the H-O-O-H torsion angle in the above example is set to 100.0°, but only this single parameter need be altered to adjust the angle. In a description of the molecule using cartesian coordinates, the x, y and z coordinates of at least one of the atoms would need to be altered, which would be very much less convenient.

Useful quantities:

$$\text{The tetrahedral angle} = 2 \cos^{-1} \frac{1}{\sqrt{3}} = 109.47°$$

Typical bond lengths:

C—H 1.07 Å   C—C 1.54 Å   C=C 1.34 Å   C≡C 1.21 Å

A.9 AMINO ACIDS

Name	3-Letter code	1-Letter code	Sidechain
Glycine	Gly	G	H
Alanine	Ala	A	Me
Valine	Val	V	$(CH_3)_2CH$
Leucine	Leu	L	$(CH_3)_2CHCH_2$
Isoleucine	Ile	I	$CH_3CH_2CH(CH_3)$
Phenylalanine	Phe	F	$PhCH_2$
Tyrosine	Tyr	Y	$HOC_6H_4CH_2$
Tryptophan	Trp	W	Indole-CH_2
Serine	Ser	S	$HOCH_2$
Threonine	Thr	T	$CH_3CH(OH)$
Methionine	Met	M	$CH_3SCH_2CH_2$
Cysteine	Cys	C	$HSCH_2$
Asparagine	Asn	N	H_2NCOCH_2
Glutamine	Gln	Q	$H_2NCOCH_2CH_2$
Proline	Pro	P	$—CH_2CH_2CH_2$
Aspartic acid	Asp	D	$HOOCCH_2$
Glutamic acid	Glu	E	$HOOCCH_2CH_2$
Lysine	Lys	K	$H_2NCH_2CH_2CH_2CH_2$
Arginine	Arg	R	$H_2NC(=NH)NHCH_2CH_2CH_2$
Histidine	His	H	Imidazole-CH_2

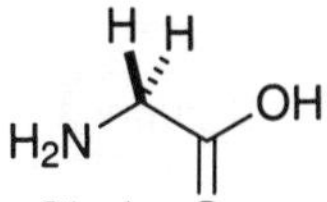

Glycine

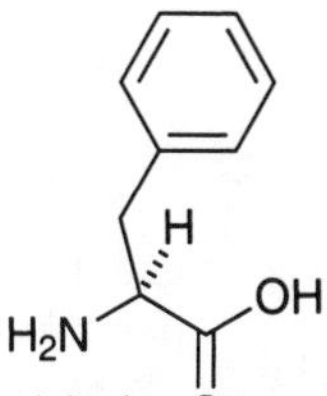

Phenylalanine

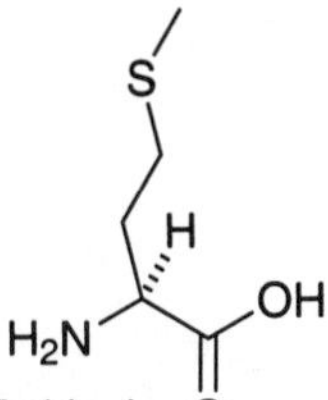

Methionine

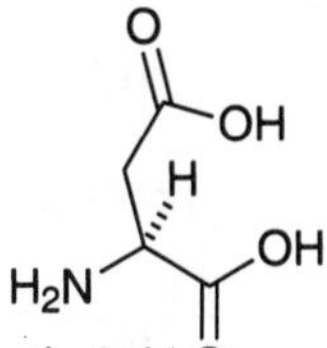

Aspartic Acid

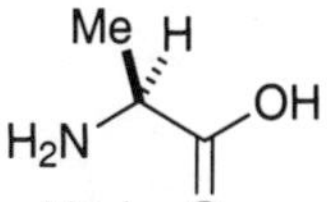

Alanine

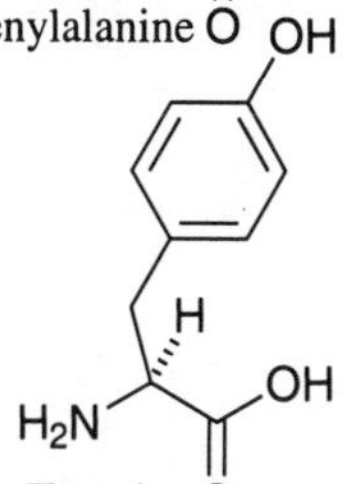

Tyrosine

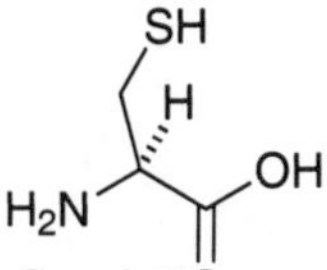

Cysteine

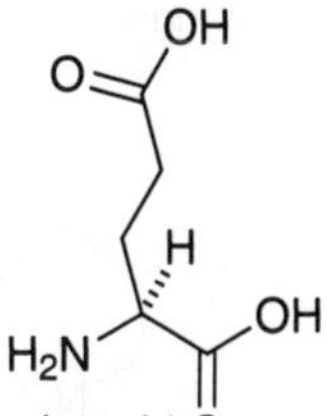

Glutamic Acid

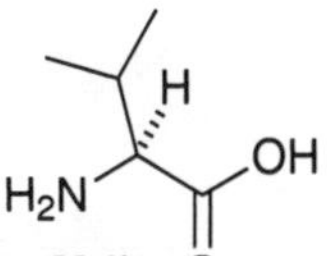

Valine

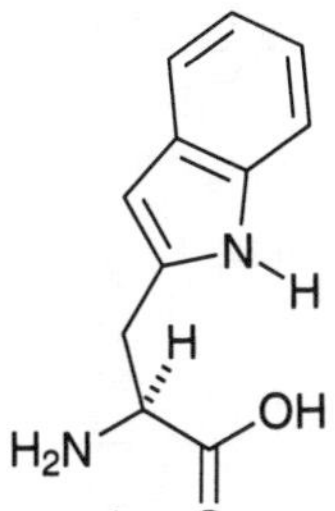

Tryptophan

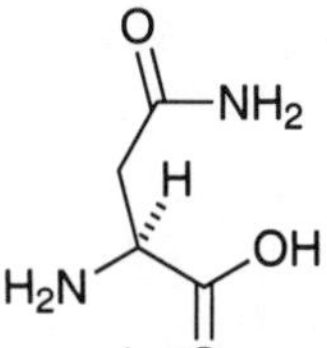

Asparagine

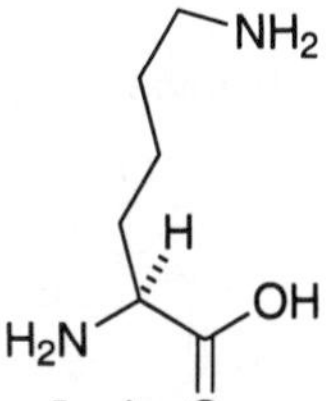

Lysine

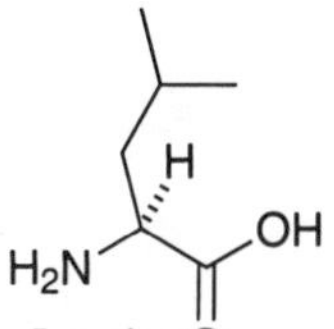

Leucine

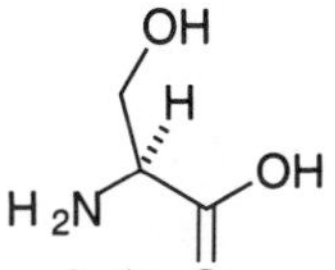

Serine

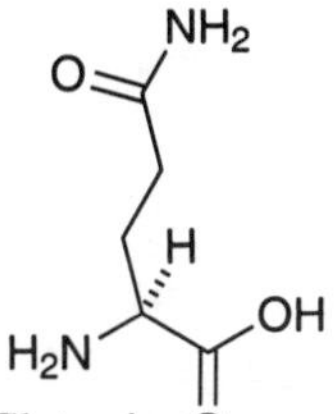

Glutamine

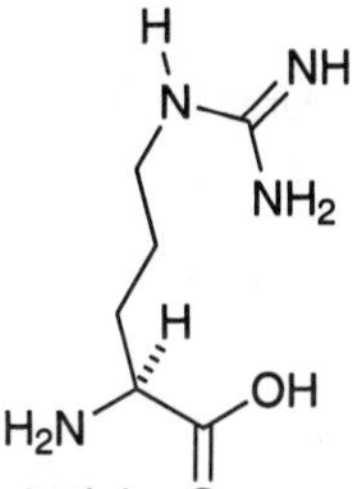

Arginine

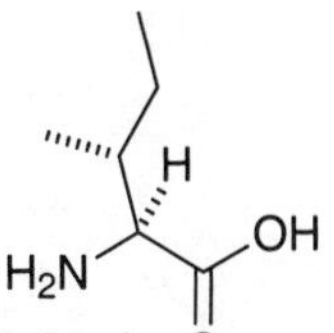

Isoleucine

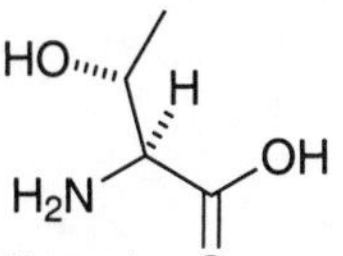

Threonine

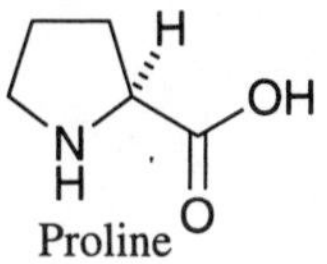

Proline

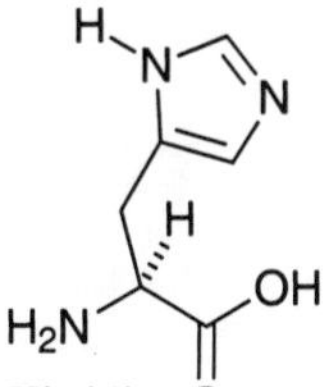

Histidine

A.10 NUCLEIC ACIDS

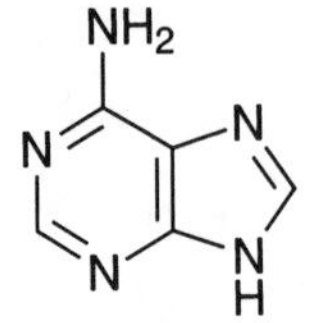

Adenine

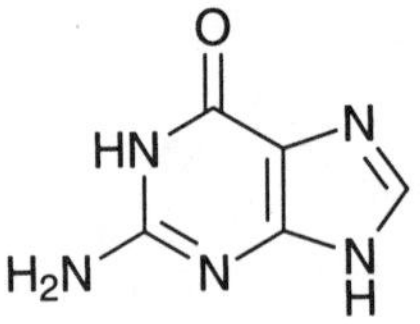

Guanine

purines

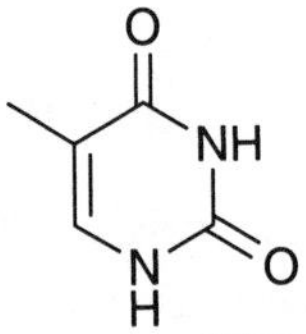

Thymine

DNA only

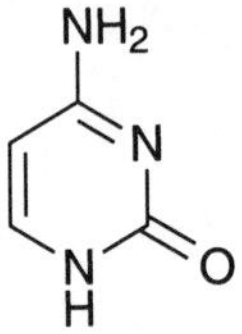

Cytosine

Uracil

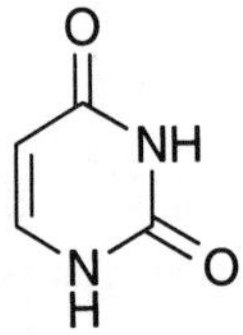

RNA only

pyrimidines

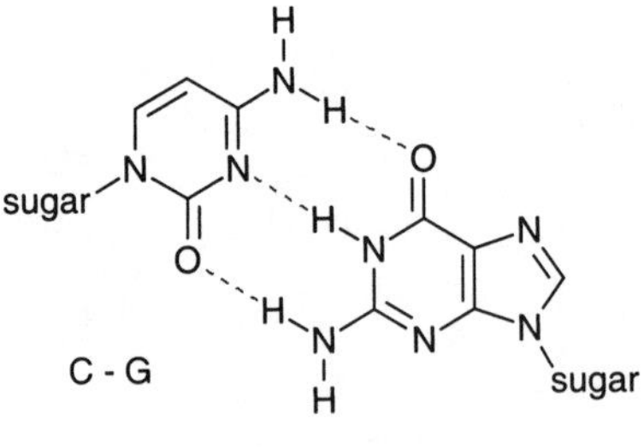

C - G

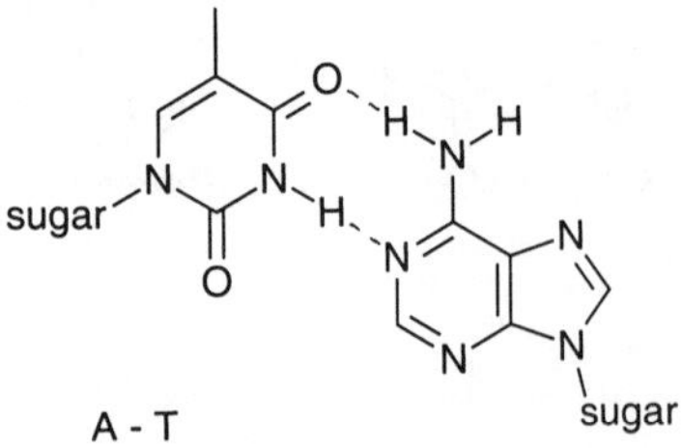

A - U

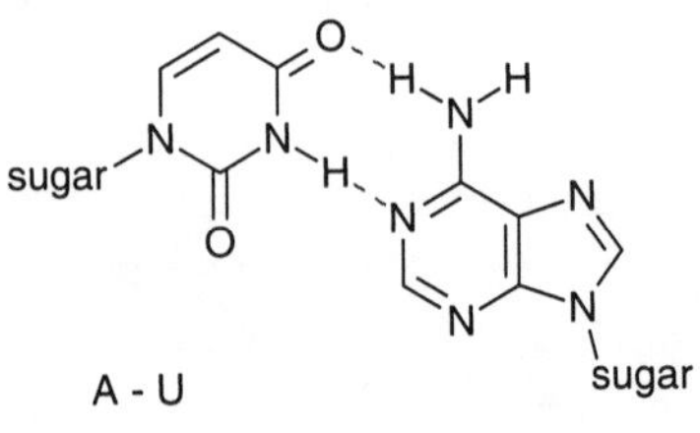

A - T

DNA

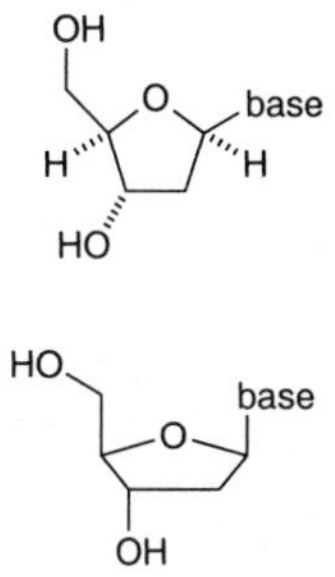

RNA

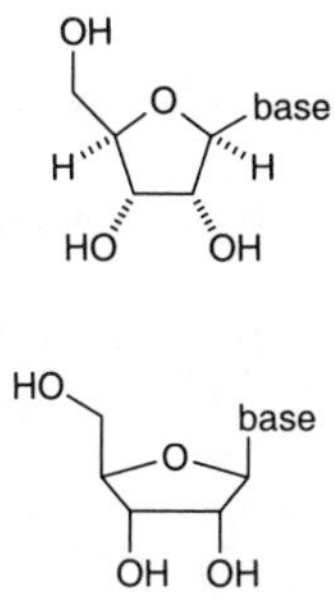

strand of DNA

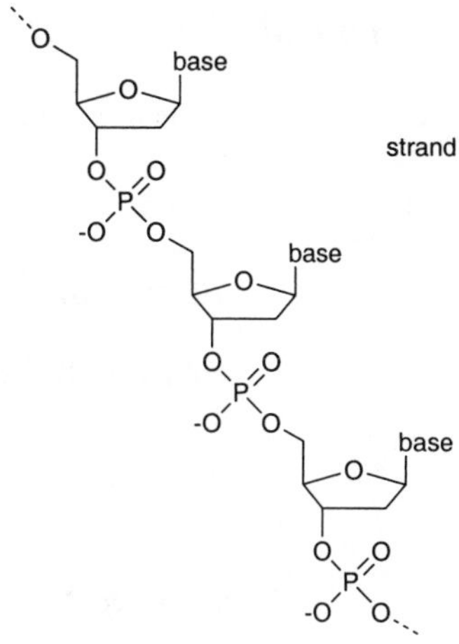

A.11 USEFUL MATHEMATICS

Taylor Series:

$$f(x + a) = f(a) + x\,f'(a) + \frac{x^2}{2!}f''(a) + \frac{x^3}{3!}f'''(a) + \frac{x^4}{4!}f''''(a) + \ldots + \frac{x^n}{n!}f^{(n)}(a) + \ldots$$

where $f'(x)$ is the first derivative of $f(x)$, $f''(x)$ is the second derivative, and $f^{(n)}(x)$ is the nth derivative.

It follows that:

$$e^x = 1 + x + \frac{x^2}{2!} + \frac{x^3}{3!} + \frac{x^4}{4!} + \ldots + \frac{x^n}{n!} + \ldots$$

$$\cos x = 1 - \frac{x^2}{2!} + \frac{x^4}{4!} - \frac{x^6}{6!} + \frac{x^8}{8!} - \ldots$$

$$\sin x = x - \frac{x^3}{3!} + \frac{x^5}{5!} - \frac{x^7}{7!} + \frac{x^9}{9!} - \ldots$$

Logarithms

$$\text{If } x = \log y, \text{ then } y = 10^x$$

This is a base-10 logarithm. It is often convenient to use e instead of 10 as a base (e $\approx$ 2.718281828). These are called natural logarithms, written ln instead of log.

$$\text{If } x = \ln y, \text{ then } y = e^x$$

Stirling's Formula:

$$\ln n! \approx n \ln n - n \text{ (best for large } n)$$

Trigonometric Formulae:

$$\sin (A + B) = \sin A \cos B + \cos A \sin B$$

$$\cos (A + B) = \cos A \cos B - \sin A \sin B$$

$$\sin A + \sin B = 2 \sin\left(\frac{A + B}{2}\right) \cos\left(\frac{A - B}{2}\right)$$

$$\cos A + \cos B = 2 \cos\left(\frac{A + B}{2}\right) \cos\left(\frac{A - B}{2}\right)$$

Integral:

$$\int_{-\infty}^{\infty} e^{-ax^2} dx = \sqrt{\frac{\pi}{a}}$$

Sine and Cosine Rules:

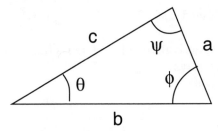

$$\text{sine rule:} \frac{a}{\sin \theta} = \frac{b}{\sin \psi} = \frac{c}{\sin \phi}$$

$$\text{cosine rule: } a^2 = b^2 + c^2 - 2bc \cos \theta$$

Vector Manipulation:

For two vectors $\mathbf{a} = (x_a, y_a, z_a)$ and $\mathbf{b} = (x_b, y_b, z_b)$

$$\text{The length of } \mathbf{a} = |\mathbf{a}| = \sqrt{x_a^2 + y_a^2 + z_a^2}$$

Dot product:

$$\mathbf{a} \cdot \mathbf{b} = |\mathbf{a}| |\mathbf{b}| \cos \theta = x_a x_b + y_a y_b + z_a z_b$$

where θ is the angle between the vectors

Vector product:
$$\mathbf{a} \wedge \mathbf{b} = (y_a z_b - z_a y_b, z_a x_b - x_a z_b, y_a x_b - y_a x_b)$$

This is a vector perpendicular to both $\mathbf{a}$ and $\mathbf{b}$, with magnitude $|\mathbf{a} \wedge \mathbf{b}| = |\mathbf{a}| |\mathbf{b}| \sin \theta$

The dot product and the vector product can be used to find the bond angle between three atoms, and the torsion angle between four atoms.

Matrix Determinants:
The determinant of an n by n matrix can be expressed in terms of n determinants of $(n-1)$ by $(n-1)$ matrices:

$$\begin{vmatrix} x_{11} & x_{21} & x_{31} & \ldots & x_{n1} \\ x_{12} & x_{22} & x_{32} & \ldots & x_{n2} \\ \ldots & \ldots & \ldots & \ldots & \ldots \\ x_{1n} & x_{2n} & x_{3n} & \ldots & x_{nn} \end{vmatrix}$$

$$= x_{11} \begin{vmatrix} x_{22} & x_{32} & \ldots & x_{n2} \\ \ldots & \ldots & \ldots & \ldots \\ x_{2n} & x_{3n} & \ldots & \end{vmatrix} - x_{21} \begin{vmatrix} x_{12} & x_{32} & \ldots & x_{n2} \\ \ldots & \ldots & \ldots & \ldots \\ x_{1n} & x_{3n} & \ldots & x_{nn} \end{vmatrix} + x_{31} \begin{vmatrix} x_{12} & x_{22} & \ldots & x_{n2} \\ \ldots & \ldots & \ldots & \ldots \\ x_{1n} & x_{2n} & \ldots & x_{nn} \end{vmatrix} -$$

$$\ldots + \ldots x_{n1} \begin{vmatrix} x_{12} & x_{22} & x_{32} & \ldots \\ \ldots & \ldots & \ldots & \ldots \\ x_{1n} & x_{2n} & x_{3n} & \ldots \end{vmatrix}$$

Applying this procedure repeatedly, the determinant of an arbitrarily large matrix can be written in terms of the sum of many determinants of small matrices. Thus for a 3 by 3 matrix:

$$\begin{vmatrix} x_{11} & x_{21} & x_{31} \\ x_{12} & x_{22} & x_{32} \\ x_{13} & x_{23} & x_{33} \end{vmatrix} = x_{11} \begin{vmatrix} x_{22} & x_{32} \\ x_{23} & x_{33} \end{vmatrix} - x_{21} \begin{vmatrix} x_{12} & x_{32} \\ x_{13} & x_{33} \end{vmatrix} + x_{31} \begin{vmatrix} x_{12} & x_{22} \\ x_{13} & x_{23} \end{vmatrix}$$

$$= x_{11}(x_{22}x_{33} - x_{32}x_{23}) - x_{21}(x_{12}x_{33} - x_{32}x_{13}) + x_{31}(x_{12}x_{23} - x_{22}x_{13})$$

A.12 ADDING PARAMETERS TO A FORCE FIELD

If a particular molecule cannot be described by any of the force fields which are available, then it may be necessary to add new parameters to an existing force field (Hopfinger and Pearlstein, 1984). This is not a trivial operation, and great care must be taken to validate the new parameters against all the experimental data which are available. The parameters which are needed are: bond lengths and strengths, bond angles and bending terms, van der Waals interactions, charge–charge interactions and torsion parameters. These are described in detail in

Chapter 2, although some force fields will need additional crossterms, which will make adding new parameters rather harder. These parameters are sufficient to describe most 'organic' molecules. Molecular mechanics descriptions of transition metals exist, but may require extra terms. If a force field is simply required to give a qualitative impression of a molecule's structure, then it may be reasonable to introduce severe constraints into the description of a metal, which will force it to behave in a limited way. If a general description of a heavy element is required, then molecular mechanics may not be able to provide a good solution.

Bond lengths and bonds angles may be found from crystallographic data, such as that stored in the Cambridge Crystallographic Database. The Brookhaven Protein Database should not be used for this purpose, because many of the crystal structures within it will already have been optimised using a force field, and so data extracted from this source may simply reproduce an existing force field. Molecular orbital calculations are also a good source of bond length and bond angle data. However, a bond length calculated by high level *ab initio* molecular orbital theory will not be precisely the same as a bond length measured from a high-quality X-ray crystallographic study, because the X-ray data give an average structure, as the molecules in crystals vibrate slightly. However, these differences tend to be smaller than the other uncertainties in force fields.

Bond stretching and angle bending parameters can be directly calculated from *ab initio* molecular orbital theory, or estimated by comparing the lengths of similar bonds which are stretched or compressed within a molecule. The best values may be found by trial and error. The units which force fields use for bond stretching are not always intuitive, and must be treated with care. Values can also be found from infra red spectroscopy, provided a particular vibration corresponds to a particular bond or angle movement. Vibrational frequencies calculated from Hartree–Fock *ab initio* molecular orbital theory tend to be too large by about 12%, and need to be scaled.

$$\text{Bond stretching constant} = k_1 = 5.892 \times 10^{-7}\, \mu\, v^2$$

where k_1 is measured in mdyn Å^{-1}, the reduced mass, μ, is measured in atomic units and the frequency, v, is measured in wavenumbers (cm^{-1}).

$$\mu = \frac{m_1 m_2}{m_1 + m_2}$$

$$\text{Angle bending constant} = k_\theta = 5.892 \times 10^{-7}\, I\, v^2$$

where k_θ is measured in mdyn Å rad^{-2}, the reduced moment of inertia, I, is measured in atomic mass units and Å^2, and the frequency, v, is measured in wavenumbers (cm^{-1})

$$I = \frac{m_1 r_1^2 m_2 r_2^2}{m_1 r_1^2 + m_2 r_2^2}$$

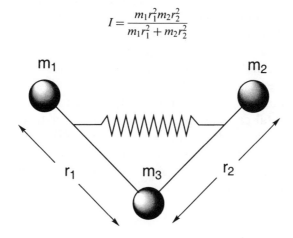

If new values for atomic size and hardness are required, it may be possible to estimate reasonable results by interpolating from the adjacent elements in the periodic table. Values for the charge distribution are hard to obtain, and molecular orbital theory is probably the best approach. Mulliken and Löwdin charges can be used for atomic partial charges, but these tend to give values which are too large (Mulliken, 1955; Löwdin, 1955). A better approach is to use electrostatic fitting to obtain the atom-centred point charges which best reproduce the electrostatic potential around the molecule, as calculated by molecular orbital theory (Breneman and Wiberg, 1990). An advantage of this method is that a fairly low level of molecular orbital theory, even semi-empirical calculations, can give good results.

Torsion parameters (usually V_1, V_2 and V_3) are particularly hard, because some of the torsion terms can be expressed by the van der Waals, bond stretching and angle bending terms. A process sometimes called *scription* (Goodman, Kahn and Paterson, 1990) can be helpful. First, the torsion parameters V_1, V_2 and V_3 are all set to zero, and energies for the structure are obtained from this incomplete force field, by constraining the torsion angle in 30° increments, and so getting an energy profile for twisting around the bond. This is fitted to a truncated Fourier series to give V_1^{MM2}, V_2^{MM2} and V_3^{MM2}. The same process is then repeated with molecular orbital calculations to produce V_1^{MO}, V_2^{MO} and V_3^{MO}. The differences between these terms are then incorporated into the force field.

New parameters should be compared with existing parameters in the force field, as a sanity check. If the new parameters are of a different order of magnitude to existing parameters, then something is likely to be wrong. It is unlikely that one bond angle is ten times stiffer than another, or a rotation barrier for a new bond is very much larger than that required to twist around a carbon–carbon double bond.

Derivation of parameters is not a deterministic process. Parameters may be best obtained by trial and error in some cases. The important thing is that there should be experimental data against which new parameters can be tested, and so validate a guessed value, or show the need for another trial set of values.

It is desirable to have as much experimental data and as many molecular orbital calculations as possible in order to test a new force field against a range of data. The final tweaking of the force field, in order to reproduce these data as well as possible, and to balance the different inaccuracies against each other is a very difficult task, and gives an idea of the likely accuracy of the final results.

References and Further Reading

FURTHER READING

Molecular Modelling

Burkert, U.; Allinger, N. L. *Molecular Mechanics*, American Chemical Society, Washington DC, 1982.

Grant, G. H.; Richards, W. G. *Computational Chemistry*, Oxford University Press, Oxford, 1995.

Hehre, W. J.; Burke, L. D.; Shusterman, A. J.; Pietro, W. J. *Experiments in Computational Chemistry*, Wavefunction Inc., 1993.

Leach, A. R. *Molecular Modelling: Principles and Applications*, Longman, Harlow, 1996.

Lipkowitz, K. B.; Boyd, D. B. *Reviews in Computational Chemistry*, VCH, New York, Eleven volumes 1990–1997.

Vinter, J. G.; Gardner, M. *Molecular Modelling and Drug Design*, Macmillan, London, 1994.

General Physical Chemistry

Atkins, P. W. *Physical Chemistry*, Oxford University Press, Oxford, 1990.

Hirst, D. M. *A Computational Approach to Chemistry*, Blackwell Scientific Publications, Oxford, 1990.

Isaacs, N. S. *Physical Organic Chemistry*, Longman, Harlow, 1995.

Molecular Dynamics

Brooks III, C. L.; Karplus, M.; Pettitt, B. M. 'Proteins: A theoretical perspective of dynamics, structure and thermodynamics' *Adv. Chem. Phys.*, Wiley, **71**, 1988.

'Molecular Mechanics and Modeling', E. R. Davidson (editor), *Chem. Rev.*, 1993, **93**, #7.

van Gunsteren, W. F.; Berendsen, H. J. C. 'Computer simulation of molecular dynamics: methodology, applications and perspectives in chemistry' *Angew. Chem. Int. Ed. Engl.*, 1990, **29**, 992–1023.

Molecular Orbital Theory

Hehre, W. J.; Radom, L.; Schleyer, P. von R.; Pople, J. A. Ab Initio *Molecular Orbital Theory*, John Wiley & Sons, New York, 1986.

Fleming, I. *Frontier Orbitals and Organic Chemical Reactions*, Wiley, New York, 1986.

Drug Design

Dean, P. M. *Molecular Foundations of Drug–Receptor Interactions*, Cambridge University Press, Cambridge, 1987.

DATA SOURCE

CRC Handbook of Chemistry and Physics, D. R. Lide (editor), CRC Press, New York, 1997.

REFERENCES

Abraham, R. J.; Rossetti, Z. L. 'Rotational Isomerism. Part XV. The Solvent Dependence of the Conformational Equilibria in *trans*-1,2- and *trans*-1,4-Dihalogenocyclohexanes' *J. Chem. Soc. Perkin Trans. 2*, 1973, 582–587.
Allen, F. H.; Kennard, O. '3D Search and Research Using the Cambridge Structural Database' *Chemical Design Automation News*, 1993, **8**, 1, 31–37.
Allinger, N. L. 'Conformational Analysis. 130. MM2. A Hydrocarbon Force Field Utilizing V_1 and V_2 Torsional Terms' *J. Am. Chem. Soc.* 1977, **99**, 8127–8134.
Allinger, N. L.; Yuh, Y. H.; Lii, J. H. 'Molecular Mechanics. The MM3 Force Field for Hydrocarbons. 1' *J. Am. Chem. Soc.* 1989, **111**, 8551–8556.
Allinger, N. L.; Chen, K.; Lii, J. H. 'An Improved Force Field (MM4) for Saturated Hydrocarbons' *J. Comp. Chem.* 1996, **17**, 642–668.
Allured, V. S.; Kelly, C. M.; Landis, C. R. 'SHAPES Empirical Force Field: New Treatment of Angular Potentials and Its Application to Square-Planar Transition-Metal Complexes' *J. Am. Chem. Soc.* 1991, **113**, 1–12.
Bader, R. F. W. 'Atoms in Molecules' *Acc. Chem. Res.* 1985, **18**, 9–15.
Baker, J. 'An Algorithm for the Location of Transition States' *J. Comp. Chem.* 1986, **7**, 385–396.
Bartell, L. S. 'On the Effects of Intramolecular van der Waals Forces' *J. Chem. Phys.* 1960, **32**, 827–831.
Barton, D. H. R. 'Interactions between Non-bonded Atoms, and the Structure of cis-Decalin' *J. Chem. Soc.* 1948, 340–342.
Begley, M. J.; Pattenden, G.; Robertson, G. M. 'Synthetic Radical Chemistry. Total Synthesis of ($\pm$)-Isoamijiol' *J. Chem. Soc. Perkin Trans. 1*, 1988, 1085–1094.
Bernardi, A.; Capelli, A. M.; Gennari, C.; Goodman, J. M.; Paterson, I. 'Transition-state modeling of the aldol reaction of boron enolates: a force field approach' *J. Org. Chem.* 1990, **55**, 3576–3581.
Bernardi, A.; Gennari, C.; Goodman, J. M.; Paterson, I. 'The Rational Design and Systematic Analysis of Asymmetric Aldol Reactions Using Enol Borinates: Applications of Transition State Computer Modelling' *Tetrahedron Asymmetry*, 1995, **6**, 2613–2636.
Bernstein, F. C.; Koetzle, T. F.; Williams, G. J. B.; Meyer, E. F. Jr.; Brice, M. D.; Rodgers, J. R.; Kennard, O.; Shimanouchi, T.; Tasumi, M. 'The Protein Data Bank: a Computer-based Archival File for Macromolecular Structures' *J. Mol. Biol.* 1977, **112**, 535–542.
Biali, S. E. 'Axial Monoalkyl Cyclohexanes' *J. Org. Chem.* 1992, **57**, 2979–2980.
Bingham, R. C.; Dewar, M. J. S.; Lo, D. H. 'Ground States of Molecules. XXV. MINDO/3. An Improved Version of the MINDO Semi-empirical SCF-MO Method' *J. Am. Chem. Soc.* 1975, **97**, 1285–1293.
Blondel, A.; Karplus, M. 'New Formulation for Derivatives of Torsion Angles and Improper Torsion Angles in Molecular Mechanics: Elimination of Singularities' *J. Comp. Chem.* 1996, **17**, 1132–1141.

Böhm, H.-J. 'The computer program LUDI: A new method for the *de novo* design of enzyme inhibitors' *J. Comp.-Aided Mol. Des.* 1992, **6**, 61–78.

Böhm, H.-J. 'On the use of LUDI to search the fine chemicals directory for ligands' *J. Comp.-Aided Mol. Des.* 1994, **8**, 623–632.

Böhm, H.-J.; Klebe, G. 'What Can We Learn from Molecular Recognition in Protein–Ligand Complexes for the Design of New Drugs?' *Angew. Chem. Int. Ed. Engl.* 1996, **35**, 2588–2614.

Bolin, J. T.; Filman, D. J.; Matthews, D. A.; Hamlin, R. C.; Kraut, J. 'Crystal Structures of *Escherichia coli* and *Lactobacillus casei* Dihydrofolate Reductase Refined at 1.7 Å Resolution' *J. Biol. Chem.* 1982, **257**, 13650–13652.

Bondi, A. 'van der Waals Volumes and Radii' *J. Phys. Chem.* 1964, **68**, 441–451.

Bredt, J. 'Steric hindrance in the bridge ring (Bredt's rule) and the *meso-trans*-position in condensed ring systems of the hexamethylenes' *Liebig Annalen der Chemie* 1924, **437**, 1–13.

Breneman, C. M.; Wiberg, K. B. 'Determining Atom-Centered Monopoles from Molecular Electrostatic Potentials. The Need for High Sampling Density in Formamide Conformational Analysis' *J. Comp. Chem.* 1990, **11**, 361–373.

Brooks, B. R.; Brucoleri, R. E.; Olafson, B. D.; States, D. J.; Swaminathan, S.; Karplus, M. 'CHARMM: A Program for Macromolecular Energy, Minimization and Dynamics Calculations' *J. Comp. Chem.* 1983, **4**, 187–217.

Brooks III, C. L.; Karplus, M. 'Solvent Effects on Protein Motion and Protein Effects on Solvent Motion – Dynamics of the Active-Site Region of Lysozyme' *J. Mol. Biol.* 1989, **208**, 159–181.

Brown, F. K.; Houk, K. N. 'Torsional and Steric Control of Stereoselectivity in Isodicyclopentadiene Cycloadditions' *J. Am. Chem. Soc.* 1985, **107**, 1971–1978.

Brünger, A. T. 'Refinement of three-dimensional structures of proteins and nucleic acids', in *Topics in Molecular Biology*, J. Goodfellow (editor), Macmillan, London, 1990, 137–178.

Brünger, A. T.; Karplus, M. 'Molecular dynamics simulations with experimental restraints' *Acc. Chem. Res.* 1991, **24**, 54–61.

Chang, G.; Guida, W. C.; Still, W. C. 'An Internal Coordinate Monte Carlo Method for Searching Conformational Space' *J. Am. Chem. Soc.* 1989, **111**, 4379–4386.

Chirlian, L. E.; Francl, M. M. 'Atomic Charges Derived from Electrostatic Potentials: A Detailed Study' *J. Comp. Chem.* 1987, **8**, 894–905.

Clark, M.; Cramer III, R. D.; Van Opdenbosch, N. 'Validation of the General Purpose Tripos 5.2 Force Field' *J. Comp. Chem.* 1989, **10**, 982–1012.

Cohen, N. C.; Blaney, J. M.; Humblet, C.; Gund, P.; Barry, D. C. 'Molecular Modelling Software and Methods for Medicinal Chemistry' *J. Med. Chem.* 1990, **33**, 883–894.

Cornell, W. D.; Cieplak, P.; Bayly, C. I.; Gould, I. R.; Merz Jr., K. M.; Ferguson, D. M.; Spellmeyer, D. C.; Fox, T.; Caldwell, J. W.; Kollman, P. A. 'A Second Generation Force Field for the Simulation of Proteins, Nucleic Acids, and Organic Molecules' *J. Am. Chem. Soc.* 1995, **117**, 5179–5197.

Crippen, G. M. *Distance Geometry and Conformational Calculations*, Chemometrics Research Studies Series 1. Wiley, New York, 1981.

Cundari, T. R.; Sisterhen, L. L.; Stylianopoulos, C. 'Molecular Modeling of Vanadium Complexes' *Inorg. Chem.* 1997, **36**, 4029–4034.

Dang, L. X.; Kollman, P. A. 'Free Energy of Association of the 18-Crown-6:K^+ Complex in Water: A Molecular Dynamics Simulation' *J. Am. Chem. Soc.* 1990, **112**, 5716–5720.

Dean, P. M. *Molecular Foundations of Drug-Receptor Interaction*, Cambridge University Press, Cambridge, 1987.

Dean, P. M.; Barakat, M. T.; Todorov, N. P. 'Optimization of Combinatoric Problems in Structure Generation for Drug Design' *New Perspectives in Drug Design*, 1995, Academic Press, San Diego.

Dewar, M. J. S. *The Molecular Orbital Theory of Organic Chemistry*, McGraw Hill, New York, 1969.

Dewar, M. J. S.; Thiel, W. 'Ground States of Molecules. 38. The MNDO Method. Approximations and Parameters' *J. Am. Chem. Soc.* 1977, **99**, 4899–4907.

Dewar, M. J. S.; Zoebisch, E. G.; Healy, E. F.; Stewart, J. J. P. 'AM1: A New General Purpose Quantum Mechanical Molecular Model' *J. Am. Chem. Soc.* 1985, **107**, 3902–3909.

Dewar, M. J. S.; Jie, C.; Yu, J. 'SAM1; The First of a New Series of General Purpose Quantum Mechanical Molecular Models' *Tetrahedron* 1993, **49**, 5003–5038.

Dirac, P. A. M. 'Quantum Mechanics of Many-Electron Systems' *Proc. Roy. Soc.* 1929, **A123**, 714–733.

Duddeck, H.; Feuerhelm, H.-T.; Snatzke, G. 'Rearrangements and Unusual Reductions with Sodium Hydride and Methyl Iodide' *Tetrahedron Lett.* 1979, **20**, 829–830.

Eksterowicz, J. E.; Houk, K. N. 'Transition-State Modeling with Empirical Force Fields' *Chem. Rev.* 1993, **93**, 2439–2461.

Ferenczy, G. G.; Reynolds, C. A.; Richards, W. G. 'Semiempirical AM1 Electrostatic Potentials and AM1 Electrostatic Derived Charges: A Comparison with *Ab Initio* values' *J. Comp. Chem.* 1990, **11**, 159–169.

Ferguson, D. M.; Raber, D. J. 'A New Approach to Probing Conformational Space with Molecular Mechanics: Random Incremental Pulse Search' *J. Am. Chem. Soc.* 1989, **111**, 4371–4378.

Frisch, H. L.; Wasserman, E. 'Chemical Topology' *J. Am. Chem. Soc.* 1961, **83**, 3789–3795.

Gasteiger, J.; Marsili, M. 'Iterative Partial Equalization of Orbital Electronegativity – a Rapid Access to Atomic Charges' *Tetrahedron* 1980, **36**, 3219–3288.

Gennari, C.; Hewkin, C. T.; Molinari, F.; Bernardi, A.; Comotti, A.; Goodman, J. M.; Paterson, I. 'The rational design of highly stereoselective boron enolates using transition state computer modeling: A novel asymmetric anti-aldol reaction for ketones' *J. Org. Chem.* 1992, **57**, 5173–5177.

Gillet, V. J.; Myatt, G.; Zsoldos, Z.; Johnson, A. P. 'SPROUT, HIPPO and CAESA: Tools for *de novo* structure generation and estimation of synthetic accessibility' *Perspect. Drug Discov. Design* 1995, **3**, 34–50.

Go, N.; Noguti, T.; Nishikawa, T. 'Dynamics of a globular protein in terms of low-frequency vibrational modes' *Proc. Nat. Acad. Sci.* 1985, **80**, 3696–3700.

Goodford, P. J. 'A Computational Procedure for Determining Energetically Favorable Binding Sites on Biologically Important Molecules' *J. Med. Chem.* 1985, **28**, 849–857.

Goodman, J. M.; Kahn, S. D.; Paterson, I. 'Theoretical studies of aldol stereoselectivity: the development of a force field model for enol borinates and the investigation of chiral enolate π-face selectivity' *J. Org. Chem.* 1990, **55**, 3295–3303.

Goodman, J. M. 'Molecular Orbital Calculations on $R^1R^2C{=}O{\cdot}H_2BF$ Complexes: Anomeric Stabilisation and Conformational Preferences' *Tetrahedron Lett.* 1992, **33**, 7219–7222.

Goodman, J. M. 'What is the Longest Unbranched Alkane with a Linear Global Minimum Conformation?' *J. Chem. Inf. Comput. Sci.* 1997, **37**, 876–878.

Goodman, J. M.; Bueno Saz, A. 'Rapid conformation searching. Part 1.Diastereomeric compounds' *J. Chem. Soc., Perkin Trans. 2*, 1997, 1201–1204.

Goodman, J. M.; Leach, A. G. 'Rapid conformation searching. Part 2. Similar compounds' *J. Chem. Soc., Perkin Trans. 2*, 1997, 1205–1208.

Goto, H.; Osawa, E. 'Corner Flapping: A Simple and Fast Algorithm for Exhaustive Generation of Ring Conformations' *J. Am. Chem. Soc.* 1989, **111**, 8950–8951.

Goto, H.; Osawa, E. 'An Efficient Algorithm for Searching Low-energy Conformers of Cyclic and Acyclic Molecules' *J. Chem. Soc. Perkin Trans. 2*, 1993, 187–198.

Groves, P.; Searle, M. S.; Waltho, J. P.; Williams, D. H. 'Asymmetry in the Structure of Glycopeptide Antibiotic Dimers: NMR Studies of the Ristocetin A Complex with a Bacterial Cell Wall Analogue' *J. Am. Chem. Soc.* 1995, **117**, 7958–7964.

Guella, G.; Chiasera, G.; Mancini, I.; Öztunç, A.; Pietra, F. 'Twelve-Membered O-Bridged Cyclic Ethers of Red Seaweeds in the Genus *Laurencia* Exist in Solution as Slowly Interconverting Conformers' *Chem. Eur. J.* 1997, **3**, 1223–1231.

Halgren, T. A. 'Merck Molecular Force Field. I. Basis, Form, Scope, Parameterization, and Performance of MMFF94' *J. Comp. Chem.* 1996, **17**, 490–519.

Hancock, R. D.; Martell, A. E. 'Ligand Design for Selective Complexation of Metal Ions in Aqueous Solution' *Chem. Rev.* 1989, **89**, 1875–1914.

Hendrickson, J. B. 'Molecular Geometry. I. Machine Computation of the Common Rings' *J. Am. Chem. Soc.* 1961, **83**, 4537–4547.

Herrmann, F.; Suhai, S. 'Energy Minimisation of Peptide Analogues Using Genetic Algorithms' *J. Comp. Chem.* 1995, **16**, 1434–1444.

Hill, T. L. 'On Steric Effects' *J. Chem. Phys.* 1946, **14**, 465.

Hill, T. L. 'Steric Effects. I. Van der Waals Potential Energy Curves' *J. Chem. Phys.* 1948, **16**, 399–404.

Holland, J. H. 'Adaptation in Natural and Artificial Systems: an Introductory Analysis with Applications to Biology, Control and Artificial Intelligence' University of Michigan Press, Ann Arbor, 1975.

Hopfinger, A. J.; Pearlstein, R. A. 'Molecular Mechanics Force-Field Parameterization Procedures' *J. Comp. Chem.* 1984, **5**, 486–499.

Houri, A. F.; Xu, Z.; Cogan, D. A.; Hoveyda, A. H. 'Cascade Catalysis in Synthesis. An Enantioselective Route to Sch 38516 (and Fluvirucin B_1) Aglycon Macrolactam' *J. Am. Chem. Soc.* 1995, **117**, 2943–2944.

Howard A. E.; Kollman. P. A. 'An Analysis of Current Methodologies for Conformational Searching of Complex Molecules' *J. Med. Chem.* 1988, **31**, 1669–1675.

Hückel, E. 'Quantum-theoretical contributions to the benzene problem. I. The electron configuration of benzene and related compounds' *Z. Physik* 1931, **70**, 204–286.

Hunter, C. A.; Sanders, J. K. M. 'The Nature of π–π Interactions' *J. Am. Chem. Soc.* 1990, **112**, 5525–5534.

Hwang, M. J.; Stockfish, T. P.; Hagler, A. T. 'Derivation of Class II Force Fields. 2. Derivation and Characterization of a Class II Force Field, CFF93, for the Alkyl Functional Group and Alkane Molecules' *J. Am. Chem. Soc.* 1994, **116**, 2515–2525.

Jorgensen, W. L.; Chandrasekhar, J.; Madura, J. D.; Impey, R. W.; Klein, M. L. 'Comparison of Simple Potential Functions for Simulating Liquid Water' *J. Chem. Phys.* 1983, **79**, 926–935.

Jorgensen, W. L.; Tirado-Rives, J. 'The OPLS Potential Functions for Proteins. Energy Minimizations for Crystals of Cyclic Peptides and Crambin' *J. Am. Chem. Soc.* 1988, **110**, 1657–1666.

Jorgensen, W. L. 'Free energy calculations: a breakthrough for modelling organic chemistry in solution' *Acc. Chem. Res.* 1989, **22**, 184–189.

Kalberer, F.; Schmid, K.; Schmid, H. 'Die Reversibilität der para-*Claisen*-Umlagerung. Zur Kenntnis der *Claisen*-Umlagerung VI' *Helv. Chim. Acta* 1956, **39**, 555–563.

Karplus, M.; Petsko, G. A. 'Molecular Dynamics Simulations in Biology' *Nature* 1990, **347**, 631–639.

Kirkpatrick, S.; Gelatt Jr, C. D.; Vecchi, M. P. 'Optimisation by simulated annealing' *Science* 1983, **220**, 671–680.

Klunder, J. M.; Caron, M.; Uchiyama, M.; Sharpless, K. B. 'Chlorohydroxylation of olefins with peroxides and titanium tetrachloride' *J. Org. Chem.* 1985, **50**, 912–915.

Kolossváry, I.; Guida, W. C. 'Comprehensive Conformational Analysis of the Four- to Twelve-Membered Ring Cycloalkanes: Identification of the Complete Set of Interconversion Pathways on the MM2 Potential Energy Hypersurface' *J. Am. Chem. Soc.* 1993, **115**, 2107–2119.

Kolossváry, I.; Guida, W. C. 'Low Mode Search. An Efficient, Automated Computational Method for Conformational Analysis: Application to Cyclic and Acyclic Alkanes and Cyclic Peptides' *J. Am. Chem. Soc.* 1996, **118**, 5011–5019.

Koopmans, T. C. 'Ueber die Zuordaung von Wellenfunktionen und Eigenwerten zu den einzelnen Elektronen eines Atoms' *Physica* 1934, **1**, 104–113.

Kuntz, I. D.; Blaney, J. M.; Oatley, S. J.; Langridge, R.; Ferrin, T. E. 'A Geometric Approach to Macromolecule–Ligand Interactions' *J. Mol. Biol.* 1982, **161**, 269–288.

Kuntz, I. D. 'Structure-Based Strategies for Drug Design and Discovery' *Science*, 1992, **257**, 1078–1082.

Lauri, G.; Bartlett, P. A. 'CAVEAT – a program to facilitate the design of organic molecules' *J. Comp.-Aided Mol. Des.* 1994, **8**, 51–66.

Leach, A. R.; Prout, K.; Dolata, D. P. 'Automated Conformational Analysis- Algorithms for the efficient construction of low-energy conformations' *J. Comp.-Aided Mol. Des.* 1990, **4**, 271–282.

Lewis, R. A.; Leach, A. R. 'Current Methods for Site-Directed Structure Generation' *J. Comp.-Aided Mol. Des.* 1994, **8**, 467–475.

Li, Y.; Paddon-Row, M. N.; Houk, K. N. 'Transition Structures of Aldol Reactions' *J. Am. Chem. Soc.* 1988, **110**, 3684–3686.

Lifson, S.; Hagler, A. T.; Dauber, P. 'Consistent Force Field Studies of Intermolecular Forces in Hydrogen-Bonded Crystals. 1. Carboxylic Acids, Amides, and the $C=O\cdots H-$ Hydrogen Bonds' *J. Am. Chem. Soc.* 1979, **101**, 5111–5121.

Lipkowitz, K. B.; Peterson, M. A. 'Molecular Mechanics in Organic Synthesis' *Chem. Rev.* 1993, **93**, 2463–2486.

Lipkowitz, K. B. 'Abuses of Molecular Mechanics' *J. Chem. Ed.* 1995, **72**, 1070–1075.

Lipton, M.; Still, W. C. 'The Multiple Minimum Problem in Molecular Modelling. Tree Searching Internal Coordinate Conformation Space' *J. Comp. Chem.* 1988, **9**, 343–355.

Löwdin, P. O. 'Quantum Theory of Many-Particle Systems. I. Physical Interpretations by Means of Density Matrices, Natural Spin-Orbitals, and Convergence Problems in the Method of Configurational Interaction' *Phys. Rev.* 1955, **97**, 1474–1489.

McCammon, J. A.; Gelim, B.; Karplus, M. 'Dynamics of folded proteins' *Nature* 1977, **267**, 585–590

McGarrah, D. B.; Judson, R. S. 'Analysis of the Genetic Algorithm Method of Molecular Conformation Determination' *J. Comp. Chem.* 1993, **14**, 1385–1395.

Makino, S.; Kuntz, I. D. 'Automated Flexible Ligand Docking Method and its Application to Database Search' *J. Comp. Chem.* 1997, **18**, 1812–1825.

Matsumoto, T.; Matsunaga, N.; Kanai, A.; Aoyama, T.; Shiori, T.; Osawa, E. 'Epimerization of Tilivalline' *Tetrahedron*, 1994, **50**, 9781–9788.

Matsumoto, T.; Aoyama, T.; Shiori, T. 'Reaction Pathway for the Epimerization of Tilivalline' *Tetrahedron Lett.* 1996, **37**, 13521–13524.

Mayo, S. L.; Olafson, B. D.; Goddard III, W. A. 'DREIDING: A Generic Force Field for Molecular Simulations' *J. Phys. Chem.* 1990, **94**, 8897–8909.

Menger, F. M.; Sherrod, M. J. 'Origin of High Predictive Capabilities in Transition-State Modeling' *J. Am. Chem. Soc.* 1990, **112**, 8071–8075.

Meza, J. C.; Judson, R. S.; Faulkner, T. R.; Treasurywala, A. M. 'A Comparison of a Direct Search Method and a Genetic Algorithm for Conformational Searching' *J. Comp. Chem.* 1996, **17**, 1142–1151.

Mohamedi, F.; Richards, N. G. J.; Guida, W. C.; Liskamp, R.; Lipton, M.; Caufield, C.; Chang, G.; Hendrickson, T.; Still, W. C. 'MacroModel – an Integrated Software System for Modeling Organic and Bioorganic Molecules using Molecular Mechanics' *J. Comp. Chem.* 1990, **11**, 440–467.

Momany, F. A.; McGuire, R. F.; Burgess, A. W.; Scheraga, H. A. 'Energy Parameters in Polypeptides. VII. Geometric Parameters, Partial Atomic Charges, Nonbonded Interactions, Hydrogen Bond Interactions, and Intrinsic Torsional Potentials for the Naturally Occurring Amino Acids' *J. Phys. Chem.* 1975, **22**, 2361–2381.

Morley, S. D.; Abraham, R. J.; Haworth, I. S.; Jackson, D. E.; Saunders, M. R.; Vinter. J. G. 'COSMIC(90): An improved molecular mechanics treatment of hydrocarbons and conjugated systems' *J Comp.-Aided. Mol. Design* 1991, **5**, 475–504.

Mujica, M. T.; Afonso, M. M.; Galindo, A.; Palenzuela, J. A. 'Enantioselective Synthesis of Cyclic Ethers via Hetero Diels–Alder Reaction. Formal Synthesis of (+)- and (−)-Laurencin' *Synlett* 1996, 983–984.

Mulliken, R. S. 'Electronic Population Analysis on LCAO-MO Molecular Wave Functions. I' *J. Chem. Phys.* 1955, **23**, 1833–1840.

Murray-Rust, P.; Rzepa, H. A.; Whitaker, B. J. 'The World Wide Web as a chemical information tool' *Chem. Soc. Rev.* 1997, **26**, 1–10.

Nair, N.; Goodman, J. M. 'Genetic Algorithms in Conformational Analysis' *J. Chem. Inf. Comput. Sci.* 1998, **38**, 317–320.

Nilges, M.; Habazettl, J.; Brünger, A. T.; Holak, T. A. 'Relaxation matrix refinement of the solution structure of squash trypsin inhibitor' *J. Mol. Biol.* 1991, **219**, 499–510.

Orr, W. J. C.; Butler, J. A. V. 'The Rate of Diffusion of Deuterium Hydroxide in Water' *J. Chem. Soc.* 1935, 1273–1277.

Parr, R. G.; Yang, W. 'Density-Functional Theory of the Electronic Structure of Molecules' *Annu. Rev. Phys. Chem.* 1995, **46**, 701–728.

Paterson, I.; Cumming, J. G.; Smith, J. D.; Ward, R. A. 'Studies in Marine Macrolide Synthesis: Boron and Silicon-Mediated Coupling Strategies for Swinholide A' *Tetrahedron Lett.* 1994, **35**, 441–444.

Pauling, L. *The Nature of the Chemical Bond*, Oxford University Press, Oxford, 1939 (Second edition 1950).

Pearlman, R. S.; Rusinko, A.; Skell, J. M.; Balducci, R. "CONCORD", Tripos Associates Inc, St Louis, MO, 1987.

Pople, J. A.; Santry, D. P.; Segal, G. A. 'Approximate Self-Consistent Molecular Orbital Theory. I. Invariant Procedures' *J. Chem. Phys*, 1965, **43**, S129–S135.

Pople, J. A.; Segal, G. A. 'Approximate Self-Consistent Molecular Orbital Theory. II. Calculations with Complete Neglect of Overlap' *J. Chem. Phys*, 1965, **43**, S136–S149.

Pople, J. A.; Beveridge, D. L. *Approximate Molecular Orbital Theory*, McGraw Hill, New York, 1970.

Pople, J. A. 'Some Deficiencies of MINDO/3 Semiempirical Theory' *J. Am. Chem. Soc.* 1975, **97**, 5306–5308.

Press, W. H.; Flannery, B. P.; Teukolsky, S. A.; Vetterling, W. T. *Numerical Recipes* Cambridge University Press, Cambridge, 1992.

Raimondi, L.; Brown, F. K.; Gonzalez, J.; Houk, K. N. 'Empirical Force-Field Models for the Transition States of Intramolecular Diels–Alder Reactions Based upon *ab initio* Transition Structures.' *J. Am. Chem. Soc.* 1992, **114**, 4796–4804.

Rao, S. N.; Kollman, P. A. 'Simulations of B-DNA. Molecular dynamics of d(CGCGAATTCGCG)$_2$ and d(GCGCGCGCGC)$_2$: An analysis of the role of initial geometry and a comparison of united and all atom models' *Biopolymers* 1990, **29**, 517–532.

Rappé, A. K.; Casewit, C. J.; Colwell, K. S.; Goddard III, W. A.; Skiff, W. M. 'UFF, a Full Periodic Table Force Field for Molecular Mechanics and Molecular Dynamics Simulations' *J. Am. Chem. Soc.* 1992, **114**, 10024–10035.

Rappé, A. K.; Colwell, K. S.; Casewit, C. J. 'Application of a Universal Force Field to Metal Complexes' *Inorg. Chem.* 1993, **32**, 3438–3450.

Rusinko, A.; Sheridan, R. P.; Nilakantan, R.; Haraki, K. S.; Bauman, N.; Venkataraghavan, R. 'Using CONCORD to construct a large database of three-dimensional coordinates from connection tables' *J. Chem. Inf. Comput. Sci.* 1989, **29**, 251–255.

Ryckaert, J. P. 'Special Geometrical Constraints in the Molecular Dynamics of Chain Molecules' *Mol. Phys.* 1985, **55**, 549–556.

Rzepa, H. S.; Whitaker, B. J.; Winter, M. J. 'Chemical Applications of the World-Wide-Web System' *J. Chem. Soc., Chem. Comm.* 1994, 1907–1910.

Sanderson, R. T. 'An Interpretation of Bond Lengths and a Classification of Bonds' *Science* 1951, **114**, 670–672.

Sands, R. D. 'Study of I-Strain Relief in the Intermediate When Forming Spiro Ketones from Unsymmetrical Cycloalkylidenecycloalkane, Their Dibromides and Their Pinacols' *J. Org. Chem.* 1994, **59**, 468–471.

Saunders, M. 'Stochastic Exploration of Molecular Mechanics Energy Surfaces. Hunting for the Global Minimum' *J. Am. Chem. Soc.* 1987, **109**, 3150–3152.

Shaver, A.; Ng, J. B.; Hall, D. A.; Soo Lum, B.; Posner, B. I. 'Insulin-Mimetic Peroxovanadium Complexes: Preparation and Structure of Potassium Oxodiperoxo(pyridine-2-carboxylato)vanadate(V), $K_2[VO(O_2)_2(C_5H_4NCOO)]\cdot 2H_2O$, and $K_2[VO(O_2)_2(OHC_5H_3NCOO)]\cdot 3H_2O$, and Their Reactions with Cysteine' *Inorg. Chem.* 1993, **32**, 3109–3113.

Sherrod, M. J. 'Empirically Optimized 'Transition State Models'' *Tetrahedron Lett.* 1990, **31**, 5085–5088.

Singh, U. C.; Kollman, P. A. 'An Approach to Computing Electrostatic Charges for Molecules' *J. Comp. Chem.* 1984, **5**, 129–145.

Skolnick, J.; Kolinski, A. 'Computer Simulations of Globular Protein Folding and Tertiary Structure' *Annu. Rev. Phys. Chem.* 1989, **40**, 207–235.

Slater, J. C. 'Atomic Shielding Constants' *Phys. Rev.* 1930, **36**, 57–64.

Squillacote, M.; Sheridan, R. S.; Chapman, O. L.; Anet, F. A. L. 'Spectroscopic Detection of the Twist-Boat Conformation of Cyclohexane. A Direct Measurement of the Free Energy Difference between the Chair and the Twist Boat' *J. Am. Chem. Soc.* 1975, **97**, 3244–3246.

Stewart, J. J. P. 'Optimization of Parameters for Semiempirical Methods. I. Method' *J. Comp. Chem.* 1989, **10**, 209–220.

Stewart, J. J. P. 'MOPAC: A semi-empirical molecular orbital program' *J. Comp.-Aided Mol. Des.* 1990, **4**, 1–105.

Still, W. C.; Romero, A. G. 'Model for the Polyepoxide Cyclization Route to Polyether Antibiotics' *J. Am. Chem. Soc.* 1986, **108**, 2105–2106.

Still, W. C.; Galynker, I. 'Chemical Consequences of Conformation in Macrocyclic Compounds. An Effective Approach to Remote Asymmetric Induction.' *Tetrahedron* 1981, **37**, 3981–3996.

Still, W. C.; Tempczyk, A.; Hawley, R. C.; Hendrickson, T. 'Semianalytical Treatment of Solvation for Molecular Mechanics and Dynamics' *J. Am. Chem. Soc.* 1990, **112**, 6127–6129.

Straatsma, T. P.; McCammon, J. A. 'Computational Alchemy' *Annu. Rev. Phys. Chem.* 1992, **43**, 407–435.

Takahashi, T.; Shimizu, K.; Doi, T.; Tsuji, J. 'Macroring Contraction Methodology. 4. A Novel Route to Steroid A, B, C Rings by the Transannular Diels–Alder Reaction of the 14-Membered (*E,E,E*)-Macrocyclic Triene' *J. Am. Chem. Soc.* 1988, **110**, 2674–2676.

Takahashi, T.; Sakamoto, Y.; Doi, T. 'Macroring Contraction Methodology. 6. Transannular Diels–Alder Reaction of the 14-Membered (*E,E,E*)-Trienone' *Tetrahedron Lett.* 1992, **33**, 3519–3522.

Teeter, M. M. 'Water structure of a hydrophobic protein at atomic resolution – pentagon rings of water molecules in crystals of crambin' *Proc. Natl. Acad. Sci. USA* 1984, **81**, 6014–6018.

Tirado-Rives, J.; Orozco, M.; Jorgensen, W. L. 'Molecular Dynamics Simulations of the Unfolding of Barnase in Water and 8 M Aqueous Urea' *Biochemistry* 1997, **36**, 7313–7329.

van Gunsteren, W. F.; Berendsen, H. J. C. 'Stochastic Dynamics for molecules with constraints: Brownian dynamics of *n*-alkanes' *Mol. Phys.* 1981, **44**, 69–95.

van Gunsteren, W. F.; Berendsen, H. J. C. 'Computer Simulation of Molecular Dynamics: Methodology, Applications and Perspectives in Chemistry' *Angew. Chem., Int. Ed. Engl.* 1990, **29**, 992–1023.

Vinter, J. G.; Hoffmann, H. M. R. 'Direct Observation of Chair–Boat Equilibria in Bridged Six-Membered Rings' *J. Am. Chem. Soc.* 1973, **95**, 3051–3052.

Vinter, J. G.; Hoffmann, H. M. R. 'Cycloadditions of Cyclic Allyl Cations to Furan.

Configuration and Conformational Analysis of the Resulting Bridged Six-Membered Rings. Isolation and Identification of Boat and Chair Atropisomers' *J. Am. Chem. Soc.* 1974, **96**, 5466–5478.

Vinter, J. G.; Davis, A.; Saunders, M. R. 'Strategic approaches to drug design. I. An integrated software framework for molecular modelling' *J Comp.-Aided. Mol. Des.* 1987, **1**, 31–51.

Vinter, J. G. 'Extended electron distributions applied to the molecular mechanics of some intermolecular interactions' *J Comp.-Aided. Mol. Des.* 1994, 8, 653–668.

Watson, J. D.; Crick, F. H. C. 'Molecular Structure of Nucleic Acids: A Structure for Deoxyribose Nucleic Acid' *Nature*, 1953, **171**, 737–738.

Weiner, S. J.; Kollman, P. A.; Case, D. A.; Singh, U. C.; Ghio, C.; Alagona, G.; Profeta Jr., S.; Weiner, P. 'A New Force Field for Molecular Mechanical Simulation of Nucleic Acids and Proteins' *J. Am. Chem. Soc.* 1984, **106**, 765–784.

Westheimer, F. H.; Mayer, J. E. 'The Theory of Racemization of Optically Active Derivatives of Diphenyl' *J. Chem. Phys.* 1946, **14**, 733–738.

Westheimer, F. H.; 'Calculation of the Magnitude of Steric Effects' in *Steric Effects in Organic Chemistry*, M. S. Newman (editor), Wiley, New York, 1956, 523–555.

White, D. N. J.; Bovill, M. J. 'Molecular Mechanics Calculations on Alkanes and Non-conjugated Alkenes' *J. Chem. Soc. Perkin Trans. 2*, 1977, 1610–1623.

Whitesides, G. M. 'What Will Chemistry Do in the Next Twenty Years?' *Angew Chem. Int. Ed. Engl.* 1990, **29**, 1209–1218.

Wong, C. F. 'Systematic sensitivity analyses in free energy perturbation calculations' *J. Am. Chem. Soc.* 1991, **113**, 3208–3209.

Yoon, K.; Parkin, G.; Rheingold, A. L. 'A Reinvestigation of the Molecular Structures of *cis-mer*-$MoOCl_2(PR_3)_3$: Do Bond-Stretch Isomers Really Exist?' *J. Am. Chem. Soc.* 1991, **113**, 1437–1438.

Yoon, K.; Parkin, G. 'Artificial Manipulation of Apparent Bond Lengths as Determined by Single Crystal X-ray Diffraction' *J. Am. Chem. Soc.* 1991, **113**, 8414–8418.

Zimmerman, H. E.; Traxler, M. D. 'The Stereochemistry of the Ivanov and Reformatsky Reactions. I.' *J. Am. Chem. Soc.* 1957, **79**, 1920–1923.

Zwanzig, R. W. 'High-Temperature Equation of State by a Perturbation Method. I. Nonpolar Gases' *J. Chem. Phys.* 1954, **22**, 1420–1426.

Subject Index